1994

Praise for

IT'S RAINING FROGS A[...]

"This delightful look at nature...reminds a[...]u[...] [...]p[...]i[...]y i[...] [...]i[...] hectic, fast-paced, just-do-it world—that it is more than OK, it is desirable, to be child-like and to look up at the heavens and ask why."
—*Toledo Blade*

"Dennis, who writes 'The Natural Enquirer' for *Wildlife Conservation* magazine, is a consummate researcher and gifted storyteller. Wolff's drawings are first rate....This book is must reading for the naturally curious of all ages, and it belongs in every school library."
—*Michigan Out-of-Doors*

"A notebook full of nuggets about the wonders of nature—how thunder and lightning form, bat sonar, the courtship of mayflies."
—Deborah A. Behler, Executive Editor, *Wildlife Conservation*

"Clear, engaging prose and lovely, lucid illustrations....A most enjoyable field guide to the awesome panoply of the sky."
—Stephanie Mills, author of *In Praise of Nature*

"A great new book for nature lovers....It's a delightful mix of hard science and folklore—which mix more often than you might think."
—*Lansing State Journal*

"When children start asking questions about nature, most parents will simply tell them the time-honored truth. For instance, thunder is merely God's bowling. And the northern lights are the angels playing flashlight tag, and so on. Not Dennis and Wolff....Basically they've put you parents out of business. From now on, there's no excuse if your kids don't know the right answers."   —*Oakland Press* (Michigan)

"Dennis and Wolff put together a charming 312-page journal filled with informative, sometimes humorous facts about the sky and its daily seasonal surprises. The wonderfully illustrated book is organized by season and will make heroes of parents and teachers, who will be able to explain nature's magic and the superstitions surrounding it."
—*El Paso Times*

It's Raining Frogs and Fishes

Four Seasons of Natural Phenomena and Oddities of the Sky

Jerry Dennis

drawings by

Glenn Wolff

HarperPerennial

A Division of HarperCollins*Publishers*

To Gerald and Eva Dennis,
for giving their rapacious nestlings a fine start.
—J.D.

To Gene and Patricia Wolff,
two saxophonists who fell in love.
—G.W.

Portions of this book first appeared, in slightly different form, in the following publications:

"Aerial Sex," "It's Raining Frogs and Fishes," "Taking the Heat," "Migrating Birds," and "Tough Birds" appeared in *Wildlife Conservation*.

"Moonstruck" appeared in *Wildlife Conservation* and *The Utne Reader*.

"Aurora Nights" appeared in *Country Living*.

A hardcover edition of this book was published in 1992 by HarperCollins Publishers.

HarperCollins books may be purchased for educational, business, or sales promotional use. For information please write: Special Markets Department, HarperCollins Publishers, Inc., 10 East 53rd Street, New York, NY 10022.

First HarperPerennial edition published 1993.

Designed by Helene Berinsky

The Library of Congress has catalogued the hardcover edition as follows:

Dennis, Jerry.
   It's raining frogs and fishes : four seasons of natural phenomena and oddities of the sky / by Jerry Dennis; illustrations by Glenn Wolff. — 1st ed.
      p.   cm.
   Includes bibliographical references and index.
   ISBN 0-06-016375-5 (cloth)
   1. Weather. 2. Natural history. 3. Seasons. I. Title.
QC981.D46   1992
551.6—dc20   92-52603

ISBN 0-06-092195-1 (pbk.)

93 94 95 96 97 DT/CW 10 9 8 7 6 5 4 3 2 1

Contents

v

ACKNOWLEDGMENTS

Of the many people who helped during the research, writing, and illustrating of this book, we would especially like to thank Carole Jean Simon and Gail Dennis, for their infinite patience and invaluable advice; and Aaron and Nicholas Dennis and Elizabeth and Sarah Wolff, for asking good questions and posing for countless sketches. We're grateful as well to Patrick Flynn, Sandra Carden, Laurie Davis, Janine Benyus, Hank Dempsey, and Jeanne Hanson for their encouragement and many excellent suggestions, and to Debbie Behler, executive editor of *Wildlife Conservation* magazine, for her good humor and gentle nudging.

Special thanks to Dr. William Scharf of the School of Biological Sciences at the University of Nebraska-Lincoln; Dr. E. C. Krupp, Director of the Griffith Observatory in Los Angeles; and Terry Root at the University of Michigan's School of Natural Resources, for reviewing portions of the manuscript.

INTRODUCTION

The world is a strange and wonderful place, no doubt about it, but it is sometimes easy to forget just how strange and wonderful it is until you have children to remind you. In the same way that they reinvent language as they learn to talk, children discover the world with the open-eyed astonishment of explorers setting foot on new lands. They look deeply and openly, and what they see fills them with questions.

Glenn Wolff and I have lived most of our thirty-odd years in northern Michigan, a place amply supplied with outdoor attractions. Except for Glenn's nine years in that center of natural wonders, Manhattan, and the two years I lived in Louisville, Kentucky, we have spent much of our time tramping in woods, canoeing, fishing, hunting, spying on birds and deer and snakes, and generally poking around outdoors. Although we are not trained naturalists—our formal schooling is in art and literature—we have always considered ourselves naturalists in the original sense of the word: enthusiastic amateurs, practicing a hands-on study of the natural world.

But when our children began asking some very good questions about that world—Why is the sky blue? How can a flock of birds all change directions at the same time? Why don't blue jays go south in the winter like robins? Are you sure no two snowflakes are alike?—we realized our knowledge was far from complete. Inspired by our kids, we became more curious. Perhaps because we passed so much time as children ourselves laying on our backs looking at clouds and birds and stars, many of the things we became most curious about take place in the sky, and are studied by astronomers, meteorologists, ornithologists, and entomologists. Somewhere in the process of exploring those disciplines in an effort to plug the holes in our educations, we realized we had a book to write.

Early in the nineteenth century the English poet John Keats complained that natural scientists, if given a chance, would "Unweave a rainbow." His complaint was valid. It is possible to explain too much, to analyze the beauty and mystery right out of things. Yet I will never forget watching one of those natural scientists, an accomplished ornithologist engaged in the daunting task of cataloguing the entire populations of gulls, terns, and cormorants nesting in the five Great Lakes. I spent a few days with him after he had already passed two exhausting years on the project, and saw him become spellbound with delight when hundreds of herring gulls —the same "flying rats" you see fighting among themselves for scraps at landfills—rose in keening, hovering ranks around him while he walked among their nests. It was obvious that unweaving the secrets of gulls does not necessarily diminish the enjoyment of them. It could be argued, in fact, that the unweaving *increases* the enjoyment.

We want to know the world and how it works—not to conquer it, but because the knowledge enhances our pleasure and deepens our appreciation. Someone once said that watching nature is a feast for the heart while understanding nature is a feast for the mind. Children, poets, and scientists have made it apparent that humans are hungry in both heart and mind. Watching is not enough. We crave answers.

This book is our effort to answer some of the questions kids ask, that adults would ask if we weren't quite so afraid of sounding childlike. We've tried to make the book as useful as possible, dividing it by season into four sections, so it can be used as a kind of field guide. Hopefully it will anticipate some excellent questions. Perhaps it will make heroes of a few parents. With luck it will add to the enjoyment of a most bountiful feast.

SPRING

23½° (DAY = NIGHT)

Vernal Equinox

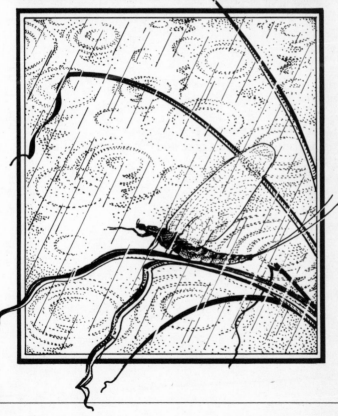

INTRODUCTION

Wherever there are four seasons, spring is synonymous with hope, youth, and fresh beginnings. The world wakes then from the deep sleep of winter and new life rises from the ground. It is the season of procreation and renewal, of vitality and fecundity. The name itself has ancient roots, and is mentioned in Old English texts as both the source of a stream and the act of leaping. As early as 1398 the word *springtime* was used in print to denote the season when the world leaps to its feet and new life springs from the ground. Later, a reference to the season as "spring of the leaf" appeared, and was subsequently shortened to "spring."

Most cultures have celebrated the end of winter as the earth's renewal or resurrection, and have marked the event with music, dancing, and feasting. Since prehistoric times people have sought to stimulate fertility by staging ritual lovemaking rites or mock battles to drive off winter. The Babylonians acted out their myths of Creation each spring. The ancient Greeks engaged in drinking, feasting, and animal sacrifice during the Festival of Flowers in honor of the god Dionysus. It evolved later into the Great Dionysia, a celebration in March that was an occasion for unrestrained revelry and the presentation of Athens' great dramas. Centuries later, the Romans took the Greeks' lead and celebrated a number of spring festivals, including the April shepherds' rite, Parilia, and the celebration of flora and sexuality during the Floralia. In the declining years of the Roman Empire, the Bacchanalia, celebrated in honor of Bacchus, the Greek and Roman god of wine and fertility, was a frenzied version of the Greek Dionysia. The Christian tradition of Lent, with its forty days of fasting and penitence from Ash Wednesday to Easter, has its sources in ancient pagan traditions of self-denial that were believed to contribute to each spring's resurrection of the earth. A Scandinavian custom pitted two opponents in mock battle, one representing winter, the other representing spring, with spring always victorious.

The English tradition of May Day probably originated in the fertility festivals of India and Egypt, and was a day of great merriment, with dancing, singing, and symbolic gathering of plants to bring luck and fertility to the community. In medieval England the celebration began early in the morning, when young adults ran to the woods to gather symbolic flowers and branches—and to perhaps engage in what one old text called "wanton dalliances." Much of the day's celebrating focused on a Maypole erected in the center of town and decorated with garlands of flowers and flowing streamers that would be held by dancers who weaved around one another as they circled the pole. Some communities replaced the Maypole with a May tree such as the white hawthorn in England or the arbutus in America, both of which were thought to possess magical powers.

Some of the fervor of those early celebrations is probably due to that widespread and well-known malady, spring fever. Modern researchers have suggested that increased energy and productivity, unexplained cheerfulness, and a tendency to break out in song are reactions to the increasing hours of sunlight in spring. More babies are conceived in the spring than during any other season, perhaps a relic holdover from a time when humans, like most mammals, may have conceived only in this season of lengthening days and growing abundance of food.

It is easy to project our own appreciation for spring onto wildlife and imagine that the frolicking of colts and the courtship rituals of birds are a celebration as well. The arrival of the birds each spring is irrefutable evidence that spring has indeed arrived. Some cultures have considered the birds not just the heralds of the change of season, but the *cause* of it. For centuries, Siberian tribes honored arriving geese by building artificial nests for them, to insure that in future years they would again return and bring spring with them.

In temperate regions, where most of the world's birds nest, each day begins with a symphony of complex and melodious bird songs. In more egocentric eras, humans liked to think birds sang strictly for our entertainment, or at least in joyous reaction to the season. Biologists usually accept less romantic explanations for avian vocalization and explain that bird song serves only to attract mates, warn off rivals, and establish feeding territories.

Most vertebrates mate in late winter or early spring. Large mammals with longer gestation periods mate in the fall to time the births of their young with spring. Spring is the most opportune season to give birth for the simple reason that for most species of animals, food is easiest to find in spring, summer, and early fall. The extra hours of daylight are critical as well, especially for birds that spend every available moment keeping demanding nestlings fed.

Spring is the transition between the extremes of winter and summer. In the northern hemisphere the vernal equinox, marking the official beginning of spring, is March 21 most years. In the southern hemisphere the spring equinox is September 22. On those days the sun passes through the celestial equator—that imaginary line across the imaginary sphere projected into the heavens around earth—and the world experiences a moment of equilibrium, with day and night everywhere on the planet equally balanced at twelve hours each. We would experience that equilibrium every day and night if the world were settled straight up and down on its axis.

The equinox is the official start of spring, but the actual beginning—the day when new plant growth rises and migrating birds return—varies from place to place. The season moves north as the sun climbs higher, progressing, according to an old rule of thumb, at the rate of about 100 miles per week. For many people, especially those in northern regions where winters are long and rigorous, spring cannot come fast enough.

Prevailing Winds

BLOW, WINDS

Blow, winds, and crack your cheeks! Rage! Blow!
You cataracts and hurricanoes, spout
Till you have drench'd our steeples, drowned the cocks!
—William Shakespeare, *King Lear*

In early March where I live, while old snow still lingers in the woods and floe ice is stacked along the Lake Michigan shore, there comes a day when the wind shifts to the south and brings the fragrant, promising odors of new growth and freshly turned soil. The wind, though not yet warm, smells as if it will soon be warm. More cold and snow may be likely in a few days or a week, but that southerly breeze is the turning point of winter. It has brought the change of seasons as surely as it will bring, in a few more weeks, Canada geese and robins and gentle rains.

As long as humans have stood on hilltops and felt the force of moving air they have wondered about its origin. In early civilizations the wind was the breath of the gods, blown gently in pleasure or tempestuously in rage. In Greek mythology it was controlled by vengeful Poseidon, the god of the seas, and by Aeolus who kept it locked for safekeeping in an enormous whistling cavern. When Aeolus played his harp men heard the music of the breeze in the trees; when he blew his conch shell great storms devastated the land and turned the ocean deadly. The four winds—Boreas, Zephyr, Notus, and Argestes—were the children of Eos, goddess of the dawn. Another goddess, Eurynome, was said to have stirred the north wind into existence by dancing, then to have mated with it and given birth to the world. According to Homer, Aeolus presented the winds tied up in a leather bag to Odysseus to aid him in his travels, but when the bag was opened by Odysseus's companions the winds escaped and whirled away to cause mischief around the world.

The Greek astronomer Anaximander was among the first in the western world to contend that the wind was not a supernatural force wielded by the gods, but a natural "flowing of air" that could be examined and studied. A century later, Anaxagoras theorized that heat caused air to rise, and that it cooled as it ascended, eventually forming clouds. Aristotle argued that the winds were dry exhalations of the sun, as opposed to the wet exhalations that caused rain. The first-century Roman naturalist Pliny the Elder offered the opinion in his *Natural History* that steady winds might fall from the stars, or result from "the continuous motion of the world and the impact of the stars traveling in the opposite direction," or come as a "breath that generates the universe by fluctuating to and fro as in a sort of womb." Gusts of wind, to Pliny, had a terrestrial origin, formed "when bodies of water breathe out a vapor that is neither condensed into mist or solidified into clouds," or were simply "the dry and parched breath from the earth."

The wind is a complex collection of forces with a simple origin: the sun. Where the sun shines longest and most directly on earth, the ground is warmed, air rises above it, and cooler air flows in to take its place. It is that flowing transfer of air from cool regions to warm that we feel as wind. In spring the slowly elevating sun moves newly heated regions northward, displacing the cold zone that had settled in during the winter. As the sun rises higher in the sky each day, newly heated air rises, creating a steep barometric gradient from north to south, allowing cold air from the north to rush in to replace the rising warm air. That air is in turn warmed, rises, and is replaced by yet more cold air.

Those same winds play an important role in melting the winter's accumulation of snow. The common belief that spring rains cause snow to disappear quickly is not true. Rain by itself, though it may make snow settle, actually melts very little snow unless it is accompanied by a warm wind. Even the sun is a slow melter of anything but dirty snow. Dirty snow absorbs a great deal of heat, every fleck of dirt and bark acting like a solar conductor into the snowpack. But it is wind that melts snow best. Warm, moist winds create condensation on the snow's surface, which in turn gives off heat and raises the temperature of the wind even more, sometimes resulting in such rapid melting that the saturated soil cannot absorb fast enough to avoid flooding. When the winds are warm and dry, like the Chinook winds of the Rocky Mountains, they can melt huge amounts of snow almost overnight, yet cause no flooding because much of the snow sublimates directly into the dry air without melting into water first.

In a less complicated world, warm air at the equator would rise, allowing cold air at the poles to flow to the equator, and all winds would blow strictly from the north or the south. Complications abound however, because air heats more quickly over land than water, as well as over certain types of land, such as asphalt parking lots, golf courses, and deserts. Uneven heating of the surface creates wind because air is ceaseless in its efforts to reach an equilibrium of temperature.

The most significant influence on global winds is the rotation of the earth itself. When the earth spins on its axis, a point of land on the equator travels at about 1,000 miles per hour to complete a revolution in twenty-four hours. But as you move toward the poles, points on the surface of the planet make progressively smaller revolutions and travel more slowly to make that same twenty-four-hour circuit. This principle, named the Coriolis effect for the French physicist who identified it in the early nineteenth century, can be demonstrated on a spinning phonograph record. The center turns at a leisurely rate to make thirty-three revolutions in one minute, while a spot on the outer rim must speed rapidly to make the same number of revolutions in the same amount of time.

Because of the Coriolis effect, when air currents travel north from the equator the ground gradually slows beneath them, causing the winds to curve to the east. Likewise, air currents traveling south toward the equator find themselves passing over ground that is continually speeding up, causing the air currents to curve toward the west. Thus storm systems in the Northern Hemisphere usually rotate counterclockwise and storm systems in the Southern Hemisphere usually rotate in a clockwise direction.

Upper atmospheric winds are predominantly westerly over much of the earth, while surface winds vary considerably from place to place. *Anabatic*, or upslope winds, are common in valleys, where air warming through the daylight hours expands and is driven uphill. After sunset, the air cools and reverses its direction, rushing back down the valley to become *katabatic*, or downslope winds.

Wind, like water descending a riverbed, is subject to friction. Hills, trees, and buildings cause land winds to be less than half the strength of corresponding winds over water. In the same way that stones in a river create surface waves, obstructions on land break the wind into gusts. Gusts are less common at sea, although they can be caused by swelling waves. At heights above about 2,000 feet the effects of surface friction can no longer be felt, and the wind blows steadily.

Anyone who has spent much time near the shore of an ocean or large lake is likely to have noticed that wind directions over the water change dramatically every day. When the weather is warm, sunlight during the day heats the land to a higher temperature than the water. As the heated air over the land rises, the cooler air above the water blows in to take its place. Daytime breezes increase as the day progresses and the land temperatures increase, reaching a peak in late afternoon. At night the situation is reversed. The water retains much of the heat it absorbed in the day, while the land quickly cools. The rising warm air over the water creates a pressure gradient that pulls cooler air toward it, and the cool land air blows out to sea. Thus, coastal breezes at night are typically offshore, blowing from the land to the water, while during the day they are usually onshore, blowing from the water to the land. Near sunset and sunrise equal temperatures on land and water will cause a temporary period of calm. Mariners in the days of square-rigged sailing vessels remembered these tendencies of the wind by saying, "In by day, out by night."

Wherever powerful and unusual winds occur they earn names

for themselves. The *blue norther* of Texas is a winter wind that precedes a fast-moving cold front, replacing warm, moist air with a furious, bone-chilling northerly wind that can drop temperatures as much as 50 degrees Fahrenheit in two or three hours.

Chinook winds, named for the Chinook Indians of the western slopes of the northern Rockies, are warm, strong, westerly winds appearing out of clear skies several times each winter in the eastern foothills of the Rockies from Colorado to Alberta, Canada. They often raise the temperature overnight by as much as 40 or 50 degrees. By the time Chinook winds reach the high plains, they have lost so much moisture in their passage over the Rockies that humidity is only about 40 percent or less. That dry, relatively warm air can evaporate snow at the rate of an inch per hour—a phenomenon that caused the Blackfoot Indians to call the wind "snow eater." One of the most dramatic of recorded Chinook winds swept down from the Black Hills to Rapid City, South Dakota, on January 22, 1943, and in the span of two minutes raised the temperature from −10 to +45 degrees Fahrenheit.

The *foehn* of the Alps is another snow eater, deriving its name from the Latin *favonius* for "south wind," or from the Gothic *fon* for "fire." A sudden hot, dry wind that sweeps down on valleys in the Alps, the foehn causes such sudden thaws that the snow often avalanches. As with the Chinook winds of North America, temperatures can rise rapidly fifty or more degrees, and the air is so hot and dry that snow is sublimated directly to water vapor. Foehn winds also dessicate the moisture from wood structures, creating fire hazards. They are caused by a low-pressure front from the northwest that cools and condenses as it rises over the mountains. On the opposite side of the summit the air that is drawn down warms about one degree Fahrenheit for every 176 feet of descent. By the time the air reaches the valley bottoms it is hot and extremely dry.

Monsoon winds, from the Arabic word *mausim* ("season"), are powerful winds that behave seasonally the way lesser coastal winds act daily. During the hot summer season in Southeast Asia, Arabia, and Australia, the heated land airs rise, sucking cooler ocean airs inland for months at a time. The winds, saturated with moisture from their passage over water, bring torrential rains with them. In winter, the winds reverse direction and the rains cease.

The mistral, or "master wind," is a sudden, harsh north wind that brings cold weather and winds of as much as ninety miles per hour down the Rhone Valley into southern France and the Mediterranean coast.

For at least forty days each year a katabatic wind known as the bora howls through the valleys of the Alps to the north coast of the Adriatic Sea. The French novelist Stendhal complained that Trieste, in 1831, was battered by high winds five days a week and the bora the other two days. "I call it a high wind," he wrote, "when I hold on to my hat, and a bora when I am in danger of breaking my arm."

The zonda is a hot, dry summer wind that sweeps down the slopes of the Andes and across the pampas in Argentina, while the tourments are "tormenting" blizzard winds that strike the same mountains in winter.

Elsewhere in the world are the buran, a powerful Russian wind that brings blizzards in winter and thunderstorms in summer; the datoo, the east wind of Gibraltar; the refreshing etesian summer breeze of Greece; frisk vind, the gale-strength wind of Sweden; the williwaw, a squall that funnels seaward through valleys in the mountains near the Strait of Magellan; and the "gyrating" land and sea winds of South America, the vrazones.

The wind is often considered a force to be feared. Ethiopian tribesmen believed evil spirits dwelt in whirlwinds and would chase them with knives. Eskimo women of the nineteenth century brandished clubs to chase the wind from their houses. The Greek historian Herodotus described Tunisians who marched into the desert with drums and cymbals to beat back a wind that had dried up their water supplies, and who perished when the wind returned and buried them in sand.

"Ill winds" were mentioned by Voltaire, who noticed a "black melancholy over the whole nation" whenever an east wind swept over France; George Eliot, who wrote, "Certain winds will make men's temper bad"; Shakespeare, who reported that north winds caused "gout, the falling evil, itch, and the ague"; and Hippocrates, the father of modern medicine, who suspected the west wind caused people to turn pale and sickly. Some modern researchers are exploring the possibility that winds can cause psychological and physical reactions in people and animals. If periods of hot, dry winds cause you to feel restless and irritable, it might be because such winds create an imbalance in the ionic content of the atmosphere. An excess of positive ions is suspected of causing the body to produce large amounts of serotonin, one of the neurotransmitter substances of the brain, which in turn cause a person to feel nervous, depressed, and irritable. Conversely, the same studies have suggested that an excess of negative ions, such as those created by falling water, cause feelings of well-being and tranquility.

Of all ill winds, few can compare in harshness to those that sweep across the deserts of the world. In Israel, the Sudan, Turkestan, the American Southwest, and other arid regions, powerful winds are usually bad news, and the source of a great deal of discomfort and danger. People of those regions treat the vast, searing-hot winds and dust-storms with respect and fear, taking cover before them, even if the only way to find cover is to lay on the ground with turbans and robes pulled over their heads to avoid breathing the hot blowing sand. Depending on where in the world you go, you might have the misfortune of running into the *sharav* of Israel, which has been accused of giving more than 25 percent of the population respiratory difficulties, headaches, nausea, and irritability; the *sirocco*, an oppressively hot, dry, dust-laden southerly wind that blows from the North African desert to Italy and the Mediterranean region, most often in spring; the *tebbad*, or "fever wind," of Turkestan; the *simoom*, or "poison wind," of Arabia and Libya, which contributes much of the dust found in the atmosphere above Europe; the *harmattan* of Algeria and Morocco, a northeastern trade wind of the western Sahara; the *khamsin* of Egypt; the *Santa Ana* of southern California; and the *haboob* dust-storms of the Sudan and the American Southwest, whose leading edges of swirling dust can tower nearly a mile high and appear as impenetrable as solid walls.

How windy can it get? North America's Great Plains, with their hundreds of miles of undulating terrain unbroken by trees or large hills, are swept by winds that average ten to twelve miles per hour,

Beaufort Scale of Wind Force

NUMBER		MPH	E · F · F · E · C · T · S
0	calm	0 ~ 1	smoke rises vertically
1	light air	1 ~ 3	smoke drifts, ripple patches on water
2	light breeze	4 ~ 7	leaves rustle, consistent water ripples
3	gentle breeze	8 ~ 12	leaves & twigs move, small waves
4	moderate breeze	13 ~ 18	small branches move, longer waves
5	fresh breeze	19 ~ 24	small trees sway, a few whitecaps
6	strong breeze	25 ~ 31	large branches sway, overall whitecaps
7	moderate gale	32 ~ 38	larger trees sway, whitecaps & swells
8	fresh gale	39 ~ 46	twigs break off trees, longer and higher waves
9	strong gale	47 ~ 54	branches break off trees, tops blow off waves
10	whole gale	55 ~ 63	trees uprooted, churning sea
11	storm	64 ~ 73	widespread damage
12	hurricane	74 ~ 136	devastation

Devised in 1805 by Sir Francis Beaufort of the British Royal Navy, the Beaufort Scale is still the standard of wind measurement used by mariners and pilots.

day after day, year-round, and frequently reach gale force or higher during winter blizzards and summer thunderstorms. Early settlers, living in sod houses in the vast expanses of the plains, joked about including a "crowbar hole" in the sides of their houses through which they could test the weather without stepping outside. If they poked a crowbar through the hole and it bent in the wind, the weather was normal; if the crowbar broke off, it was a day better spent indoors.

The most powerful wind ever recorded—exceeded only, perhaps, by the strongest winds within tornadoes—was a 231 mile-per-hour gust recorded on April 12, 1934, at the top of Mount Washington in New Hampshire. That same mountain peak experiences winds in excess of 100 miles per hour during at least one "century day" each month, and in winter sees hurricane strength, seventy-five-mile-per-hour winds nearly every day. Mount Washington and East Adelie Land, Antarctica, are probably the breeziest places in the world. The winds at both locations, day and night, year-round, average forty miles per hour.

Although we tend to think of wind as a potentially destructive force, it serves many life-giving purposes as well. It cleans the atmosphere, broadcasts seeds, moderates temperatures around the world, and distributes rains that might otherwise fall only over the oceans and coastal regions.

It also leaves its shape wherever it blows. It falls to the surface of lakes, ruffling the water in blossoms of turbulence, then pushes waves and blows their tops into whitecaps. It forms sand dunes into elegant, shifting sculptures. It twists trees on open hillsides into natural bonsai creations. On snow fields it builds ridges of *sastrugi* and etches delicate trails where it slips into eddies around stones and trees.

We think the wind is invisible, but it is not. My son, watching gusts rolling like ocean swells across a wheat field, could only shake his head in amazement at the undulating and elegant motion of the waves, and say, "So *that's* what the wind looks like."

Wind and Weather Prediction

WIND FROM THE WEST is best, at least in most parts of the world, where westerly winds predominate and indicate stable weather patterns. On the other hand, an easterly wind is usually caused by the counterclockwise rotation of a low-pressure center and is likely to precede unpleasant weather. The east wind's reputation is so bad it has found a permanent place in the world's weather lore. In New England it is commonly said, "when the wind is from the east, neither good for man nor beast." The Bible several times mentions the evil of an east wind: "God prepared a vehement east wind" (Jonah 4:8); "The east wind hath broken thee in the midst of the seas" (Ezekiel 27:26); "The east wind brought the locust" (Exodus 10:13).

A steady wind is a sign of stable weather; a shifting wind suggests change. The old proverb, "a backing wind means storms are nigh; a veering wind will clear the sky," is often accurate. To test it, imagine a compass as a clock face, with north at twelve o'clock. If a change in the wind follows a clockwise direction, say from north to east, the wind is *veering* and fair weather is likely. If the change is counterclockwise, from north to west, the wind is *backing* and you can expect bad weather.

The Sphere of Air

Taken all in all, the sky is a miraculous achievement.
—Lewis Thomas, *The Lives of a Cell*

If we could look at the earth in cross section, the thin membrane of gas and clouds enveloping it would have the relative thickness of an apple's skin. Contained within that skin are almost all the ingredients essential to life on our planet. It is where we live. Without it we could have no hope of surviving.

The atmosphere is often described in terms that make it seem like an assembly of well-lubricated parts working together to produce rain, snow, sleet, and hurricanes—a machine for the production of meteorological phenomena. Yet the atmosphere itself might be the greatest meteorological phenomenon of all. It is such a complex structure that mechanical models of how it works are always in danger of collapsing. The meteorologists who try to make sense and order of it are correct barely half the time when they try to predict weather more than two or three days in advance. Meteorology is an imprecise science, not because of faulty computers or lack of data, but because the atmosphere is subject to so many variables that it defies easy understanding.

A century and a half ago, the amateur meteorologist Luke Howard, an Englishman who would earn world fame for devising the system of cloud classification still used today, wrote, "The ocean of air in which we live and move, in which the bolt of heaven is forged, and the fructifying rain condensed, can never be to the zealous Naturalist a subject of tame and unfeeling contemplation." Zealous naturalists like Luke Howard look at the sky and can not help wondering what keeps it from seeping away into deep space,

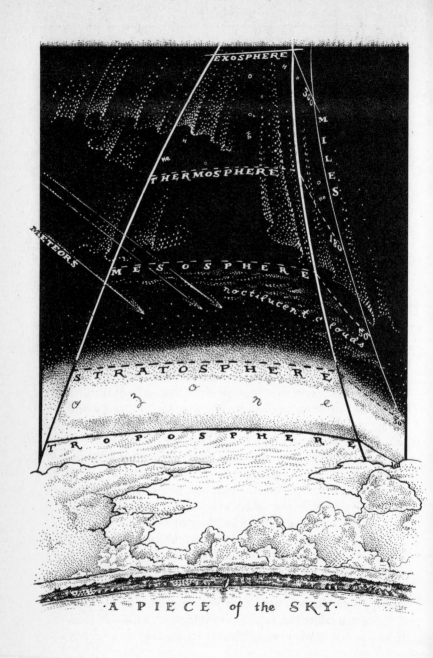

A PIECE of the SKY.

how it cleanses itself of smoke and dust and poison, why it appears blue in daylight and transparent at night. Like countless people before and after him, Luke Howard must have wondered what, exactly, this thing was that the Greeks called the "sphere of air."

In text-book definitions, atmosphere is simply a gaseous envelope contained by gravity around a planet. It can be a poisonous mixture of carbon dioxide and sulfur dioxide, as on Venus, or a seemingly bottomless cloud of methane and ammonia, as on Jupiter. Earth's atmosphere extends hundreds of miles above the ground, yet 90 percent of the air it contains is compressed within the lowest ten miles. At ground level air is composed of approximately 78 percent nitrogen, 21 percent oxygen, and 1 percent such gases as argon, carbon dioxide, helium, krypton, neon, and xenon. The amount of water vapor it contains fluctuates greatly, but on average is equal to about 2 percent of the volume of other gases. The atmosphere contains as well a rich and varying brew of dust, smoke, volcanic debris, oceanic salt particles, pollen, seeds, bacteria, and insects.

Such analysis gives little hint of the complex nature of earth's atmosphere, or of its role in the daily weather, the annual seasons, and the long-term climate of the planet. It is active in three dimensions, affected by (and affecting) both vertical and horizontal winds, influenced by the uneven heating of the sun on the earth's surface, by the motion of the earth's orbit, and by such cataclysms as volcanoes and nuclear explosions.

Significant progress in the study of the atmosphere did not begin until about the turn of the twentieth century, when the French meteorologist Léon-Philippe Teisserenc de Bort began sending up unmanned balloons carrying barometers, thermometers, and other scientific instruments. Until then the only explorations of the atmosphere had been in manned balloon flights that had reached only to about six miles, the height at which cold and insufficient oxygen made further exploration dangerous. For centuries it had been assumed that temperatures decreased steadily with altitude until the atmosphere blended with the unimaginable cold of space. Teisserenc de Bort discovered that temperatures in the atmosphere declined steadily from the ground up to about seven miles, then remained constant as high above that level as his balloons could reach. Based on those observations, he suggested in 1902 that the atmosphere be divided into layered segments. He called the lowest layer, from the ground to about seven miles up, the *troposphere*, from Greek roots meaning "sphere of change," because it is the level where temperatures vary and produce winds and clouds. The level above the troposphere, where temperatures were constant, he called the *stratosphere*, from Greek for "sphere of layers," because it was his theory that in the steady temperatures at those altitudes there would be no winds, thus gases would settle into layers, with the heavier gases arranged below the lighter ones.

In the years since, Teisserenc de Bort's observations have been refined and adjusted, and additional levels of atmosphere have been identified and studied using high-altitude balloons, aircraft, and satellites, yet his ideas about the lower reaches of the atmosphere have proven to be mostly correct. We now know the troposphere varies from about ten miles high at the equator, to about seven miles high in temperate regions, to about five miles high at the poles, and that it contains all the warm air, most of the clouds and water vapor, most of the wind, most of the oxygen and other gases, and most of the living organisms contained in the atmosphere.

When parcels of heated air from the surface rise by convection they expand and cool as they rise through the troposphere, and are almost always stopped at its top, a ceiling known as the *tropopause*. The tropopause causes the distinctive anvil-shape at the top of enormous, towering thunderheads. Temperatures in the troposphere vary from 100 degrees Fahrenheit or more at ground level, to a low of about −70 degrees Fahrenheit at the tropopause, declining at a rate of about 3.6 degrees for every 1,000 feet of altitude.

Above the tropopause, reaching to a height of about thirty miles, is the stratosphere. Temperatures throughout most of the stratosphere range from −40 degrees to −100 degrees Fahrenheit, and it is windless except for the undulating jet streams, which sometimes extend into its lower reaches.

The *mesosphere* extends from about thirty miles to about fifty miles above sea level. So few gases are found at this level that there is little to absorb solar heat. Temperatures once again plummet with altitude, reaching as low as − 225 degrees Fahrenheit at the top of the mesosphere. Paradoxically, the lowest temperatures in the mesosphere occur in summer, the warmest in winter, perhaps the result of a not-yet-understood global exchange of cold and warm air. Most meteors burn up at this level.

The next layer of atmosphere, the *thermosphere*, extends from 50 to 180 miles above the ground, and contains less than 1/100,000 of the air in the entire atmosphere. Yet even this extremely rarified layer of atmosphere absorbs a great deal of the sun's radiation— enough that temperatures fluctuate from daytime highs of 3,600 degrees Fahrenheit to nighttime lows more than 1,000 degrees cooler. Because there is no wind or convection, the thermosphere satisfies Teisserenc de Bort's original theory about the nature of the stratosphere; its gases settle into distinctive layers, with heavy nitrogen and oxygen at the bottom, and lighter helium and hydrogen at the top.

At the top of the atmosphere is the final layer, the *exosphere*, a bleak, stark region 180 to 300 or more miles high where solitary atoms of helium, hydrogen, and oxygen are likely to drift for six miles on average before colliding with other atoms or molecules. At sea level, in comparison, molecules of gas can travel only 3/1,000,000 inch before colliding. A few of those scattered atoms of air in the upper exosphere, heated by solar radiation to temperatures of 3,600 degrees Fahrenheit, speed away from the earth at 25,000 miles an hour and escape into space.

There is no clearly defined boundary between the atmosphere and space. Somewhere in the region located 300 to 900 miles above the surface of the earth, molecules are so scarce that you can safely say there are none at all.

Strong Winds Aloft:
The Jet Stream

BOMBER PILOTS FLYING at high altitudes over Japan during World War II noticed surprising and previously unsuspected ribbons of powerful wind high in the atmosphere. After the war, high-altitude balloon and jet flights revealed that these rivers of air flowed at speeds between 60 and 150 miles per hour, and occasionally reached speeds up to 290 miles per hour. They were typically 180 to 300 miles wide, one or two miles thick, and located in the upper troposphere or lower stratosphere over the subtropics and the polar regions.

It was found that the jet streams, as they came to be known, are caused by changes in air temperature when cold air from the poles moving toward the equator meets warm air from the equator moving toward the poles. The resulting winds circle the globe in several places, always from west to east, and serve to distribute heat away from the equator, helping to equalize climates, and occasionally merging to spawn massive storms. Transcontinental aircraft use the jet stream to save time and fuel by riding with them on east-bound trips and avoiding them altogether when traveling west, south, or north.

A Self-cleaning System

AS ANYONE KNOWS who has stepped outside on a bright morning after a night of rain, a good rain flushes the sky clean. Raindrops and ice crystals form around tiny specks of matter in the atmosphere, whether ash from a volcano or sulfur dioxide from a smokestack or salt from ocean spray. Once formed, a drop of rain or a snowflake continues to be an efficient cleanser, collecting other fine particles as it descends toward the ground.

Pollutants like volcanic ash, that rise high into the relatively windless and nearly moisture-free stratosphere, take much longer to be flushed from the system. Over a period of two to five years the pollutants circulate toward the planet's poles, then descend into the troposphere. Impurities there are precipitated (though rarely—it snows very little at the north and south poles) or are swept by lower-level winds to temperate regions where rain and snow flush them to the ground.

What Is the Ozone Layer,
and Why Does It Matter?

IN 1913, the French physicist Charles Fabry discovered that significant quantities of ozone were suspended in a band six to thirty miles above the ground. Ozone is produced when ultraviolet radiation divides a two-atom oxygen molecule, and the two single atoms attach themselves to other oxygen molecules, creating three-atom clusters. Those molecules are most dense between twelve and eighteen miles high in a region of the atmosphere designated as the *ozonosphere*. Even at their densest they are widely scattered: If all the ozone molecules in the atmosphere were compressed into a solid layer it would measure only $\frac{1}{10}$-inch thick.

That thin shield serves a critical purpose, however. Ozone molecules absorb ultraviolet rays, and prevent most of the sun's dangerous radiation from reaching the earth. It is not yet known precisely what effect a reduction in the ozone layer might have on life on earth, but it is certain that humans could expect increases in skin cancer and ailments of the eyes, and it is possible that such life-giving organisms as plankton in the ocean and bacteria in the soil would be destroyed by too much ultraviolet light. Even the most skeptical scientists admit that chlorofluorocarbons, used as propellants in aerosol cans and as refrigerants in air conditioners, destroy atmospheric ozone and are probably the cause in recent years of an alarming reduction in the ozone layer globally and of a virtual hole in the layer over Antarctica.

Why Is the Sky Blue?

THERE ARE NO blue skies on the moon. If you were to stand on its airless surface in daylight you would see the white sun shining brightly from a background of pure black scattered with stars. Without an atmosphere there is nothing, day or night, to obscure the view of space.

On earth, our very rich atmosphere scatters the light from the sun, bouncing it off particles of dust and molecules of gas until it is dispersed throughout the sky. Only the brightest objects are visible through that blue screen: the pale sliver of a quarter-moon, Venus after sunrise, the occasional meteor so large it sears across the daylit sky like a doomed aircraft in silent flames.

Sunlight is white, composed of an equally distributed spectrum of all the visible colors. When white light shines through a prism it separates into individual colors, each of which varies in wavelength and intensity. Red light, with its long wavelengths, is the least energetic in the visible spectrum and the least likely to be scattered by interaction with molecules in the atmosphere. Blue light has the most energy of any color we can see, and is the most intensely scattered by bouncing off the air's atoms and molecules. By the time light from the sun passes through our atmosphere and reaches the surface of the earth, much of the blue light has been filtered away from it and only the yellow and red rays reach us in a direct line, causing sunlight to appear mostly yellow to us. The blue rays of the spectrum meanwhile have scattered, zigzagging each time they encounter particles, until they spread across the entire sky and reach our eyes finally as the predominant color of the sky. Increase the scattering of light and remove yet more color —as when the sun shines horizontally through the atmosphere at sunrise and sunset, or through a sky polluted with volcanic ash or smoke from a forest fire—and the sun appears red because only the red light of the spectrum is able to resist that additional filtering.

Particles in the air scatter blue light more than other colors, causing blue to be the predominant color of the sky.

CONDENSATION NUCLEI

Warmer   Colder

water vapor   CONDENSATION   ice crystals

COALESCENCE TO SNOWFLAKES
THAT WARM TO RAINDROPS

CONDENSATION TO DROPS HEAVY ENOUGH TO FALL

COLDER LAYERS KEEP OR TURN IT BACK TO SNOW

How Comes The Rain?

CONDENSATION

HYDROLOGIC CYCLE

PRECIPITATION

EVAPORATION

SNOW   RAIN

SUBLIMATION

ICE

TRANSPIRATION

INFILTRATION

SURFACE RUNOFF

WATER TABLE

FLOW OF GROUNDWATER

It's Raining,
It's Pouring

A good rain demands celebration. Water is the most precious substance on earth, and yet it falls in abundance, freely, from the sky. We should catch it in silver bowls and tally every drop. It should inspire us, like children, to dance in circles, run stomping through puddles, and sing in gratitude.

The water that falls so freely has been here for billions of years. Its original source is a matter of speculation, but it was probably formed when hydrogen and oxygen were forced together during the condensation of gases that first formed the earth. As the planet contracted it squeezed water outward to the surface, where it settled in the low-lying areas to become the first oceans. Some water may have come also via multitudes of comets that crashed into the planet early in its history.

The earth's water remains here in a virtually constant quantity. Small amounts are lost in the upper atmosphere when water vapor is separated into its component atoms, allowing active hydrogen to slip away into space. Some water, known by geologists as *juvenile water*, continuously reaches the surface from the interior of the planet, escaping through volcanoes and other vents that reach deep beneath the crust. There is also evidence that the bombardment of protons from the sun may create small amounts of new water when the charged nuclei of hydrogen combine with free-floating oxygen in the upper atmosphere. Still, chances are good that the drop of rain that splashes on your forehead is made of molecules that were here long before the first humans looked up in wonder at a cloudy sky, long before the first leafy plants stretched their roots into the soil, long before the first single-celled organisms took the critical step of dividing in half to reproduce.

Water is a restless substance. Its molecules unlock easily, allowing it to change shape and form, to be evaporated into vapor or condensed into droplets or frozen into ice. When it appears as raindrops it is but one step in a continuous process of evaporation, condensation, and precipitation called the *hydrological cycle*. That cycle and the way water behaves in general on our planet are apparently unique. No other body in the solar system combines the necessary ingredients to maintain water in its life-giving forms. Smaller planets and moons lack the gravitational attraction to keep vapor and water on the surface from escaping into space. Hotter planets cannot keep it from boiling away; colder planets can only keep it locked up in ice caps.

Raindrops are formed in a cloud when thousands of tiny water droplets, each condensed around a microscopic bit of dust or other particle of matter that can serve as a condensation nucleus, join together into units large enough and heavy enough to fall toward the ground without evaporating. They fall, not in the classic teardrop shape so often portrayed, but flattened at the top and bottom, like tiny hamburger buns. In the tropics, where clouds even at high altitudes remain warm, the droplets begin as large raindrops, which collide with smaller ones as they fall, growing and splitting during their journey to the ground. Elsewhere in the world, precipitation almost always begins as ice crystals in the cold tops of moisture-laden clouds. The crystals grow quickly as supercooled water vapor in the cloud adheres to them, then fall as snowflakes that melt into raindrops when they descend into warmer air.

An average raindrop is 15 million times larger than a droplet in a cloud, and contains trillions of molecules of H_2O. When it falls from a moving rain cloud, the first drops pass through warm, dry air that evaporates all but the largest of them. As those few large drops strike the ground, they cool and humidify the air, reducing the evaporation of other drops and paving the way for a downpour. In extremely hot and dry conditions, such as a desert, *all* the raindrops might evaporate, creating visible curtains of rain in the air called *fallstreaks*, or "virga," which trail from beneath a cloud but disappear before they reach the ground.

Why do some places get more rain than others? In general, rainfall amounts are greatest on windward coastal slopes near warm tropical waters and smallest on the lee sides of mountains, near the center of continents, and in high latitudes. The forced lifting of winds over mountains such as those in the western United States causes ascending moist air to expand with the decreasing pressure of height until it cools and condenses, which can result in rainfall in excess of 100 inches per year in a windward area, while the lee side, in what meteorologists say is a rain "shadow," collects less than ten inches. Most of the great deserts of the world lie in a region 30 or 35 degrees on either side of the equator, where permanent high-pressure zones and erratic winds prevent the accumulation of moisture-laden air. Mariners called those same latitudes over the oceans the horse latitudes, perhaps because the frequent calms and hot weather were often fatal to horses being transported to the New World, or because the winds, when they did blow, were so erratic and unpredictable they reminded sailors of unruly herds.

In some parts of the world, when it rains it pours. The wettest spot on earth is Mount Waialeale, Hawaii, where during a recent thirty-two-year period it rained an average of 460 inches each year. On July 4, 1956, Independence Day picnics were spoiled in Unionville, Maryland, when a storm dropped 1.23 inches of rain in one minute. Rockport, West Virginia, experienced a comparable deluge on July 18, 1889, with nineteen inches of rain falling in two hours and ten minutes. The greatest single rainfall ever recorded in a twenty-four-hour period was on Réunion Island in the Indian Ocean, where it rained a total of seventy-four inches on March 15 and 16, 1952. In one incredibly damp year, from August 1860 to August 1861, the city of Cherrapunji, India, was drenched with 1,042 inches of rain.

At the other extreme, not a drop of rain fell on Arica, Chile, for more than fourteen consecutive years, from October 1903 to January 1918. The longest dry spell in the United States to date was in Bagdad, California, where 767 days passed between rains, from October 3, 1912, to November 8, 1914. Death Valley is consistently the driest spot in the United States, receiving only about 1.2 inches of rain per year. But Death Valley is like a dripping rain forest compared to the aforementioned Arica, which during a fifty-nine-year period received an average of .03 inches of precipitation each year. At that rate it would take thirty-three years to tally up an inch of precipitation, and more than fourteen centuries to equal the forty-three inches of rainfall New York City receives in one year.

People living in arid regions or who rely on agriculture for their livelihood have always been subject to the whims of rain. Because too little or too much rain can be disastrous, the weather-wise observer, magician, or shaman who can predict when the rain will come has always been highly regarded. Rainfall prediction was one of the most important duties of the shamans of Arizona's Pueblo Indians, the Masai of Africa, and other people of the world's desert regions. To the Aztecs, living in semidesert Mexico, rain was so important that Tlaloc, the god of rain and thunder, who brought life and fertility to the earth, was one of their primary deities. Tlaloc's domain was a paradise of soothing rains, where rainbows always arched, butterflies always flew, and flowers always bloomed. Much of a shaman's magic was no doubt grounded in good observation skills. He could sit on a rise of ground, notice the wisps of clouds above a distant mountain, the direction of the wind, the colors of the sunset and sunrise, and the flights of birds, and announce with fair certainty when rain could be expected.

We dance or pray or hire rainmakers to make the rain start, then recite charms to make it stop. "Rain, rain, go away, come again another day," has been chanted in various forms for hundreds of years. In 1659 it was included in a book of English proverbs as "Raine, raine, goe to Spain; faire weather come againe." Later it appeared in print as "Raine, raine, goe away, Come againe a Sater-day," and as "Rain, rain, pour down, And come na' mair to our towne." Such charms were thought to be more potent if you looked unflinchingly at a rainbow as you recited them, a trick that frequently worked since the sight of a rainbow, at least in the late afternoon or evening when the wind is from the west, is a fairly sure sign the rain is ending and clear skies are approaching.

It is probably inevitable that superstitions about rain would be passed to such a ready emblem of it as the umbrella. Many people still consider it bad luck to open an umbrella indoors, and if they drop one they will ask someone else to retrieve it rather than tempt fate by picking it up themselves. In fine weather an umbrella is meant to be carried closed. Open it and it will surely cause rain.

Smells Like Rain

AS A CHILD I was told that the peculiar, fresh, pleasant scent that always lingered in the air before a rainstorm was ozone. The theory, as I understood it, was that molecules of ozone were carried to earth with raindrops to create the distinctive fresh odor of a rain shower. Like so many of the notions we pick up as children, this one was charming but wrong. Most of the atmosphere's ozone exists far above the reach of rain clouds. Ozone, though named from the Greek word for "smell"—because of the characteristic odor it gives off when sparked during an electrical discharge—is odorless unless burned.

In ancient times it was believed that rain became sweetly scented from its passage through the heavens, and that even rainbows had an odor. In fact, however, the odor we associate with rain has a terrestrial rather than an atmospheric source. When we smell approaching rain, we are actually smelling oils that have been given off by plants and absorbed in the soil, where they blend with earthy odors. The oils, and the odors, are released into the air when the relative humidity at ground level increases to more than about 80 percent. Because humid air transmits odors more readily than dry air, we are made more receptive than usual to the heady, musky scent of the earth. And because the scent is so often followed by rain we learn at an early age to associate the two.

Rain That Kills

RAIN IS SO CRUCIAL to most forms of life on earth that it is hard to believe it can be a killer. When water vapor and raindrops mix with certain human-made pollutants—primarily sulphur dioxide and nitrogen oxide—they are chemically transformed into raindrops from Hell: sulfuric and nitric acid, which fall to earth and sicken forests, dissolve stone buildings and monuments, and make lakes and rivers incapable of supporting life.

About 10 percent of the pollutants responsible for acid rain come from such natural sources as volcanoes, forest fires, and nat-

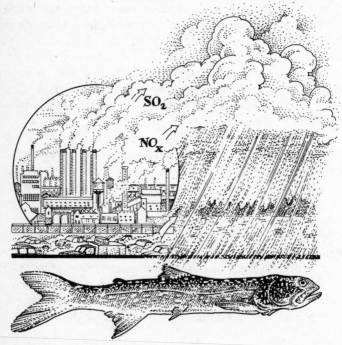

Even slight acidification of lakes and rivers can cause starvation of trout and other aquatic life.

urally decaying organic matter. The rest are pumped into the atmosphere from auto emissions, commercial and residential heating units, private industries, and electric utilities powered by coal or other fossil fuels. Although much has been done to limit emissions of those pollutants, one of the difficulties of ultimately controlling them is that they are often carried by winds hundreds and even thousands of miles before mixing with rain and falling to the ground. Lakes and rivers well hidden within the forests of northern New England and eastern Canada are suffering some of the worst acid destruction, though those regions contribute almost none of the pollutants. And while the United States, Canada, England, and other leading industrial nations are adopting stringent regulations to limit the discharge of pollutants, many developing countries continue to pump massive amounts of sulphur dioxide and nitrogen oxide into the atmosphere.

Rain, Drizzle, or Mist?

METEOROLOGISTS CLASSIFY the types of rain according to the size of raindrops and the intensity of their fall, as noted in the chart below:

	Intensity (inches/hr.)	Diameter of Drops (mm.)	Drops (sq. ft. per second)
Fog	.005	.01	6,264,000
Mist	.002	.1	2,510
Drizzle	.01	.96	14
Light Rain	.04	1.24	26
Moderate Rain	.15	1.60	46
Heavy Rain	.60	2.05	46
Excessive Rain	1.60	2.40	76
Cloudburst	4.00	2.85	113

PRIMARY BOW

ALEXANDER'S DARK BAND

SECONDARY BOW

SUN RAYS

RAINDROP

RAY BUNDLES ENTERING THE TOPS OF DROPS REFLECT COLORS AT ABOUT AN ANGLE OF 42°

REFRACTED RED TO VIOLET RAYS

42°

.02–.04 INCH

←SUN

RAIN

THE VIEWER ALWAYS FACES THE MIDDLE OF THE BOW. ENTER THE RAIN AND THE BOW VANISHES

AFTER APRIL SHOWERS: RAINBOWS

Few of the world's natural phenomena have inspired as much poetry and song as rainbows. Unlike most unusual events of the sky, rainbows have been perceived more often with wonder and delight than fear, perhaps because their appearance immediately following rain storms makes them so easily associated with optimism, cheerfulness, and good fortune. They can represent good luck—as in the pot of gold that might be found where the elusive ends of the rainbow meet the ground—or divine blessing, as in the biblical story of the deluge, when God made the rainbow a token of his covenant with Noah and his descendants: "I set my bow in the cloud, and it shall be a sign of the covenant between me and the earth. When I bring clouds over the earth and the bow is seen in the clouds, I will remember my covenant which is between me and you and every living creature of all flesh; and the waters shall never again become a flood to destroy all flesh."

Yet, not all cultures have been as enamored with rainbows as we are in the West. The Karens of Burma considered rainbows dangerous demonic spirits capable of devouring the human spirit. They blamed rainbows for sudden and violent deaths by drowning or falling, and reasoned that the hard work of taking a life made it thirsty enough to appear in the sky and drink water. Its appearance was considered a portent of yet another death to come. The Zulus of southeast Africa associated the rainbow with snakes and feared it. They believed it drank from pools of water when its ends touched the earth, and that it could inhabit those pools and seize and devour people unfortunate enough to be caught bathing there. Anyone who came in contact with a rainbow on dry land would be afflicted with disease or other misfortune. A modern supersti-

35

tion reported from North Carolina says that a house overarched by a rainbow will soon experience a death, and that a sure way to experience a death in the family within the year is to walk through the end of a rainbow.

The Kaitish tribe of central Australia saw the rainbow as more a troublemaker than as a malign spirit. In their mythology, the rainbow was the son of the rain and caused droughts when, in his duty to his father, he kept him from falling upon the earth. To prevent the rainbow from stopping rain, an elaborate ritual was enacted in which rainbows were painted on the ground, on the body of a ritual rainmaker, and on a shield. The shield would be hidden in a secure place, imprisoning the real rainbow and preventing it from interfering with the business of conjuring up a shower.

To some cultures the characteristic shape of a rainbow suggested weaponry. Tribesmen of western Siberia, Finland, and Lapland believed it was a bow used by their thunder god to shoot arrows of lightning. A similar image appears in Hindu mythology, with the god Indra firing arrows with a rainbow.

In Greek mythology the rainbow was represented by Iris, wife of Zephyrus, daughter of Electra and Thaumas, who was made to serve as a messenger between mortals and the gods. Iris dressed in multicolored robes and ran along the rainbow to deliver her messages. Our word "iridescence" is her most enduring legacy.

The Norse had a similar myth, in which the Rainbow Bridge linked the realm of the gods with the mortal world. Some Polynesian peoples believed the rainbow was not a bridge, but a ladder that their heroes could climb to reach heaven.

Australian Aborigines were ambivalent about rainbows; they saw the rainbow as a giant serpent arched across the sky. They variously describe the Rainbow Serpent as either the creator or destroyer of the world, as a spirit that lives in streams and springs and creates rain, or as a tenuous transporter of shamans on spiritual journeys across the sky.

The idea that gold and other treasures can be found at the ends of the rainbow dates back to at least the middle of the nineteenth century, according to E. C. Krupp in his fascinating study of myths and legends of the sky, *Beyond the Blue Horizon*. Jacob Grimm mentioned the legend in his *Teutonic Mythology*: "Where the rainbow touches the earth, there is a golden dish." Of course the difficulty with finding gold at the end of the rainbow is finding the end of the rainbow. Walking toward it can be a singularly fruitless exercise —it tends to move away from you as wind moves the rain clouds away. Even if you manage to catch up to the rain the rainbow proves elusive, because the refraction of light that creates the rainbow effect depends on an observer being placed between the sun and the raindrops: As soon as you enter the rain where that activity takes place, you can no longer see the reflected light and the rainbow disappears. It will seem to evade you, dimming and slipping farther into the rain as sunlight is reflected away by the raindrops behind you. The most you can hope is to get a bearing on a portion of hillside or field ahead of you where the rainbow seems to strike the ground.

Although rainbows were undoubtedly the subject of spirited debates for thousands of years, Aristotle was among the first in the Western world to give them serious analytical consideration. His interpretations of the causes and mechanics of rainbows were carefully thought out and based on the correct premise that they are reflections of sunlight on raindrops. Pliny took a more common-sensical (and less scientific) approach by declaring, with characteristic confidence, "The common occurrences we call rainbows have nothing miraculous or portentous about them, for they do not reliably portend even rain or fine weather. The obvious explanation of them is that a ray of the sun striking a hollow cloud has its point repelled and is reflected back to the sun, and that the diversified coloring is due to the mixture of clouds, fires, and air."

In 1304, long before much was known about the science of optics, a German Dominican monk named Theodoric experimented with light shining through a glass globe filled with water and proposed some theories of reflection and refraction that were astonishingly accurate. He discovered that some of the light entering his glass globe—and he assumed the globe duplicated a giant raindrop—was refracted or bent by the outer surface, reflected off the concave inner surface, then bent again as it came out. He noticed that the light exiting the globe was always only one color, but that the color changed as he moved to different positions. In what must certainly have been viewed as a crackpot conclusion in his day, Theodoric proposed that each of the colors visible in a rainbow is reflected from a different group of raindrops, and that the entire rainbow is the product of millions and millions of drops all contributing their particular color. Theodoric next applied himself to the problem of the second rainbow sometimes visible above a primary one. Noticing that the primary rainbow's sequence of colors always ended with red at the top, while the secondary rainbow always had its strip of red at the bottom, he proposed that light sometimes entered raindrops at such an angle that it skipped twice off the back of each drop, causing a double reflection. In that case, rays striking the tops of the raindrops appear as the lower, or primary, rainbow while those striking the bottoms of the drops appear as the higher, or secondary, rainbow. The secondary rainbow is dimmer than the first, Theodoric reasoned, because some of the light is lost during the double reflection, and it appears in a reversed color sequence for the same reason an image in a mirror is always reversed.

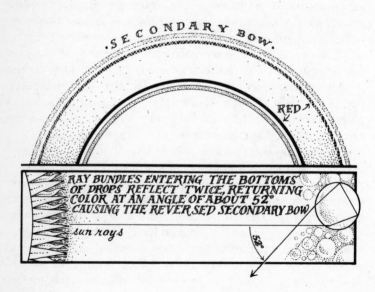

.SECONDARY BOW.

RED

RAY BUNDLES ENTERING THE BOTTOMS OF DROPS REFLECT TWICE, RETURNING COLOR AT AN ANGLE OF ABOUT 52° CAUSING THE REVERSED SECONDARY BOW

sun rays

52°

More than three centuries later, when René Descartes and Isaac Newton conducted their own experiments with light rays and glass globes, many of Theodoric's speculations proved to be accurate. Descartes confirmed that rainbows are the result of the sun's rays refracted and reflected by falling raindrops. Newton demonstrated that sunlight passing through a raindrop is sorted by wavelength, with each color sent out from the drop on a slightly different path. Thus drops high in the sky send red wavelengths to an observer, while those below send yellow, and those farther below send blue.

For years it was assumed that large raindrops produced the brightest and most vividly delineated colors, while fine drops would result in blurred colors or no colors, as is the case with fog. Recent research, however, suggests that as large drops fall they are distorted into flattened oscillating shapes that contribute little or nothing to a rainbow. Contrary to earlier theories, it is likely that the brightest colors of a rainbow are caused by tiny, nearly spherical, non-oscillating drops measuring between 2/100 and 4/100 of an inch in diameter—hardly larger than drizzle.

A rainbow bows because it is the visible portion of a perfect circle. The center of the circle, called the antisolar point, is below the horizon at a point equal to the height of the sun above the opposite horizon. At sunrise or sunset the bow will be high, a nearly complete half-circle, with its center resting on or only slightly below the horizon. When the sun is higher in the sky, the antisolar point is lower, and we see only a section of the top of the circle, causing the arc of the rainbow to appear much smaller.

Fully-formed half-circle rainbows are rare. When sections of the bow are missing it is because no rain is falling in that area or because clouds, hills, or other obstructions are blocking some of the sun's rays.

The best opportunity for viewing rainbows is immediately after a rain shower passes, when the sky behind the storm begins to break up and the sun to shine through. Typically, in most parts of the world, a rainbow will occur in late afternoon, when the prevailing winds are sweeping the storm toward the east, and the sun in the west is shining on the dark backside of the clouds. If you stand facing the receding rain, with the sun behind you, the sun-light coming over your shoulder will strike the droplets falling from the rain-bearing clouds, be separated into its component colors, and reflect back to you. The higher you stand, and the lower the sun is in the sky, the higher and more complete the arc of the rainbow will appear. Standing on the ground you will never see a complete circle—the rainbow will always be rooted in the earth. It is possible, however, to see a rainbow form in a complete circle if you are in an aircraft flying above rain-filled clouds with a high sun shining down on them.

As every child who has ever watered a lawn or garden can attest, the easiest way to summon a rainbow is with a garden hose. Press your thumb over the end of the nozzle to produce a fine, misting spray, then adjust your position until you are standing between the sun and the mist. You should be able to see at least a vivid segment of a rainbow. If you are feeling lucky, shift your position until the spray carries one leg of the rainbow to the ground. Mark the spot. Dig.

Moonbows

UNTIL ARISTOTLE PUBLISHED *Meteorologica*, his exhaustive study of weather phenomena, the notion that rainbows could occur at night was usually treated as superstition. Aristotle's scholarly opinion changed that:

> The rainbow occurs by day, and also at night, when it is due to the moon, though early thinkers did not think this ever happened. Their opinion was due to the rarity of the phenomenon, which thus escaped their observation: for though it does occur, it only does so rarely. And the reason for this is that the darkness hides the colors, and a conjunction of many other circumstances is necessary, all of which must coincide upon a single day of the

month, the day of the full moon. For it is on that day that the phenomenon must occur if it is to occur at all, and occur then only at the moon's rising or setting. So we have only met with two instances of it over a period of more than fifty years.

Much of the rarity of the phenomenon is a matter of simple logistics. The full moon rises in the east, simultaneously with the setting of the sun in the west. Most rainstorms travel from west to east in the temperate regions of the world. For a rainbow or moonbow to be visible, however, the source of light—in this case the moon—must shine fully on the rain clouds, which happens infrequently when a storm is approaching, because it is so often preceded by cloud cover. The best time to see a moonbow then is just before dawn (too early in the morning for most of us) when the setting moon is low in the western sky and can shine on the back of a receding rain cloud.

Fogbows

MORE COMMON than moonbows are colorless rainbows formed on the face of dense banks of clouds. I saw a spectacular example a few years ago in a steep river valley in Newfoundland. I was driving east, with the late afternoon sun low in the western sky, and as I descended into a valley filled nearly to the top with a seemingly impermeable river of fog, I was startled to see a perfectly formed fogbow arching from ground to ground across the face of the fog. Remarkably, the road turned straight into the bow with the sun directly behind me. The closer I approached the fogbank, the smaller and more defined the fogbow became, until it stood only about ten feet high, straddling the road from one shoulder to the other, and I seemed on the verge of driving directly beneath it, as if it were an archway into a land of clouds. But it was not possible to pass through that archway, since a fogbow or a rainbow is visible only when light strikes water droplets in front of an observer. At the moment when I seemed about to pass beneath the bow—and by now I was barely inching ahead on the highway—it dimmed, faded, and disappeared.

There is no color in a fogbow because the droplets of water in fog are so small—measuring less than $1/500$ inch in diameter—that the colors merge completely until only white is visible. The effect is strange and eerie and unforgettable, like seeing the ghost of a rainbow.

IT'S RAINING
FROGS AND FISHES

When it comes to truly peculiar weather, most of us lead sheltered lives. The chances are good that we have seen nothing more remarkable fall from the sky than snowflakes and hailstones. Naturally, we learn to expect a fair amount of order and predictability from precipitation.

But in 1921, an ichthyologist at the American Museum of Natural History challenged all notions of security by publishing a paper in the journal *Natural History* treating seriously an aspect of weather that had seldom been treated seriously before. The ichthyologist, E. W. Gudger, titled his paper "Rains of Fishes." Those who read it would never feel the same about a cloudy sky.

Gudger, an editor of the *Bibliography of Fishes*, had run across a newspaper story describing small fish found scattered on New York City streets after a heavy shower in the summer of 1824. The description intrigued him. Investigating further he uncovered a report by a woman in Indiana who claimed to have found a live minnow swimming in rainwater in a hollow in her woodpile's chopping block. Eventually Gudger collected similar accounts, ranging back hundreds of years, of people claiming to have witnessed (or nearly witnessed—many were second- or third-hand accounts) fish descending from the sky during rain storms.

In his *Natural History* paper, Gudger presented forty-four of these accounts, including ones from such widely varied sources as the second century A.D. Greek author Athenaeus who wrote in *Deipnosophists*, "I know, too, that it has rained fishes in many places . . . in Chersonesus it rained fishes for three whole days . . . certain persons have in many places seen it rain fishes, and the same thing often happens with tadpoles," and a volunteer weather observer

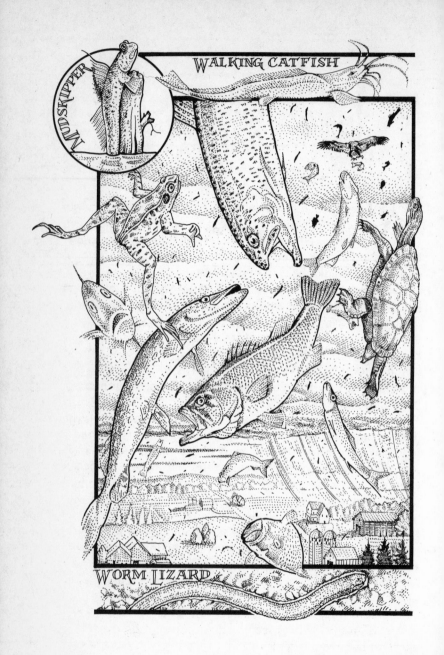

MUDSKIPPER

WALKING CATFISH

WORM LIZARD

in Tiller's Ferry, South Carolina, who reported to the June 1901 *Monthly Weather Review* that "during a heavy local rain about June 27, there fell hundreds of little fish (cat, perch, trout, etc.) that were afterwards found swimming in the pools between the cotton rows in [an adjacent] field."

Among the incidents Gudger collected were a "boisterous" rain of herring in Scotland in 1821, with the fish so abundant that tenants of the land were compelled to turn the booty over to their landlord; the depositions of ten residents of Jelalpur, India, who observed a fall of fish one noon in February 1830 and collected from the ground and from the roofs of houses large numbers of fish weighing up to six pounds; and the 1861 report of Sir J. Emerson Tennent, author of *Sketches of the Natural History of Ceylon*, who witnessed a violent rain shower on the road ahead of him and found "a multitude of small silvery fish one and one-half to two inches in length leaping on the gravel of the high road, numbers of which I collected and brought away. . . . The spot was about half a mile from the sea and entirely unconnected with any water course or pool."

Another typical report was published in the magazine *The Athenaeum*, July 17, 1841:

On Thursday week, during a heavy thunder-storm, the rain poured down in torrents mixed with half-melted ice, and, incredible as it may appear, hundreds of small fishes and frogs in great abundance descended with the torrents of rain. The fish were from half an inch to two inches long, and a few considerably larger, one weighing three ounces; some of the fish have very hard pointed spikes on their backs, and are commonly called suttlebacks. Many were picked up alive. The frogs were from the size of a horsebean to that of a garden-bean; numbers of them came down alive, and jumped away as fast as they could, but the bulk were killed by the fall on the hard pavement. We have seen some alive to-day, which appear to enjoy themselves, in a glass with water and leaves in it.

More recently, the New York Times reported, in 1931, that a rain of perch in Bordeaux was so heavy "motor cars were compelled to halt."

Convinced such evidence was incontrovertible, Gudger listed four possible explanations for the phenomenon of falling fishes. First, that witnesses had observed certain species known to migrate overland, such as mud-skippers and *Clarias batrachus*, the "walking catfish" of Asia. Second, that overflowage from ponds, streams, and gutters had left fish stranded on land. Third, that estivating fish, awakened by heavy rains, had burrowed to the surface. And fourth, that fish had been lifted from a nearby ocean or lake by waterspouts or tornados, carried overland and dumped to the ground.

Waterspouts are certainly capable of the job. Gudger noted that most of the eyewitness accounts of falling fish included mention of nearby or recent thunderstorms, monsoon rains, or waterspouts, all accompanied by strong winds. Numerous cases have been documented of powerful waterspouts lifting large objects into the air, including a five-ton houseboat at Dinner Key near Miami in 1968. A few hours after a 3,000-foot-high waterspout dispersed over Martha's Vineyard on August 19, 1896, a downpour of saltwater rained on the island. Photographs prove that waterspouts can raise large quantities of water from the surface of a lake or ocean. It is possible that a school of fish swimming near the surface could be sucked up, funneled high into the clouds by the rotating updraft, carried a few miles inland, and released with the rain as the power of the wind diminished. Theoretical calculations and measurements confirm that golfball-sized hail requires an updraft of more than 100 miles per hour, which would be more than powerful enough to loft small fish high into a thundercloud. Convinced that waterspouts or other actions of the wind were almost certainly responsible for some of the fish-falls, Gudger announced that anyone who entertained doubts about the documentation he had collected was clearly unable to "properly evaluate evidence."

One advocate of a fifth theory was Charles Fort, a renegade scientist who virtually devoted his life to searching out references to unusual natural phenomena such as rains of animals. In *The Book of the Damned*, Fort listed references to falls of a dozen species of fish and reptiles, as well as such varied stuff as fungi, stones (some inscribed with arcane messages), hatchets, masks, and masses of protoplasm. To explain such falls, Fort argued for the existence of a "Super-Sargasso Sea" suspended a few miles above the earth— "just above the reach of gravity"—in which interstellar flotsam collected and was occasionally displaced. Fort attracted a small but dedicated following—*The INFO Journal*, a magazine devoted to Fortean philosophy, is still published today—but for obvious reasons his theories were never accepted by mainstream scientists.

It was left to Gudger to alert the scientific community for the first time to the possibility that rains of creatures may have basis in fact. Until then rains of frogs and fishes were usually lumped with such subjects as teleportation, phantom lights, and the spontaneous combustion of humans. Gudger was the first scientist of undisputed reputation to seriously address the possibility that such phenomena could occur.

Not everyone was convinced. The loudest skeptic was scientist Bergen Evans, a devoted debunker of pseudoscience who in 1946 engaged Gudger in an exchange of letters in *Science* magazine debating the authenticity of Gudger's eyewitness accounts. In his book *The Natural History of Nonsense*, Bergen proposed that stories of rains of earthly creatures (and the peculiar theories of Charles Fort) have their roots in ancient mythologies and biblical references to aerial waters "above the firmament." He noted that a majority of the eyewitness accounts recorded in the past two centuries occurred during the latter part of the nineteenth century, when paleontological battles raged to explain fossilized remains of marine animals on mountaintops. Leakage from an airborne ocean was one theory proposed during that debate.

In spite of Evans's skepticism, some accounts of falling creatures are difficult to dispute. A. D. Bajkov, a biologist for the Department of Wildlife and Fisheries in Louisiana, witnessed a fall on October 23, 1947, in Marksville, Louisiana, as he reported soon after in *Science* magazine:

In the morning of that day, between seven and eight o'clock, fish ranging from two to nine inches in length fell on the trees and in yards, mystifying the citizens of that southern town [Marksville]. I was in the restaurant with my wife having breakfast, when the waitress informed us that fish were falling from the sky. We went immediately to collect some of the fish. The people in town were excited. The director of the Marksville Bank, J. M. Barham, said he had discovered upon arising from bed that fish had fallen by hundreds in his yard, and in the adjacent yard of Mrs. J. W. Joffrion. The cashier of the same bank, J. E. Gremillion, and two merchants, E. A. Blanchard and J. M. Brouillette, were struck by falling fish as they walked toward their places of business about 7:45 a.m. There were spots on Main Street, in the vicinity of the bank (a half block from the restaurant) averaging one fish per square yard. Automobiles and trucks were running over them. Fish also fell on the roofs of houses.

They were freshwater fish native to local waters, and belonged to the following species: large-mouth black bass, goggle-eye, and hickory shad. The latter species were the most common. I personally collected from Main Street and several yards on Monroe Street, a large jar of perfect specimens, and preserved them in Formalin, in order to distribute them among various museums . . .

Suddenly umbrellas seem woefully inadequate. Just when you think it is safe to run around singing in the rain, you get struck on the head by a hickory shad. As Mark Twain was fond of saying, if you aren't happy with the weather, wait five minutes. Who knows what the rain will bring?

Other Bizarre Rains

FISH ARE NOT the only living creatures reported to have fallen with the rain. Athenaeus, nearly 1,800 years ago, repeated this rather chilling account:

In Paeonia and Dardania it rained frogs, and so great was their number that they filled the houses and streets. Well, during the first days the people killed them and shut up their houses and made the best of it. But soon they could do nothing to stop it; their vessels were filled with frogs, they were found boiled or baked with their food. Besides, they could not use the water, nor could they set foot on the ground amidst the heaps of frogs piled up, and being overcome also with disgust at the smell of the dead creatures, they fled the country.

Pliny the Elder's *Natural History* listed a number of inexplicable rains:

It is entered in the records that in the consulship of Manius Acilius and Gaius Porcius [114 B.C.] it rained milk and blood, and that frequently on other occasions there it has rained flesh . . . and that none of the flesh left unplundered by birds of prey went bad; and similarly that it rained iron in the district of Lucania . . . the shape of the iron that fell resembled sponges. But in the consulship of Lucius Paullus and Gaius Marcellus [49 B.C.] it rained wool. . . . It is recorded in the annals of that year that while Milo was pleading a case in court it rained baked bricks.

Numerous other reports of rain containing frogs, toads, and other animals have filtered into the press over the centuries. Snails were said to fall on Chester, Pennsylvania, in 1870, and in 1953 a rain containing "hundreds of thousands" of snails was said to have fallen on Algiers. During a rain in 1954, crayfish were scattered across parts of Florida. In 1896 hundreds of dead ducks, wood-peckers, and catbirds fell on Baton Rouge, Louisiana. Spiders fell during a heavy rain over Hungary in 1922. Maggots—hundreds of thousands of maggots—rained on Acapulco in 1968. The *New York Post* reported that a falling clam struck a boy on the shoulder during a rainstorm in Yuma, Arizona, in 1941.

Eyewitnesses claimed a hailstone containing two living frogs fell near Dubuque, Iowa, on June 16, 1882 (the frogs hopped away after the ice melted), and that a hefty hailstone containing a six-inch gopher turtle fell in 1894 near Vicksburg, Mississippi.

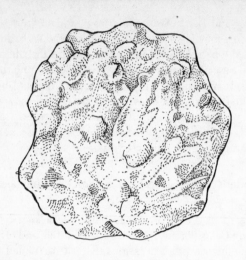

Symon's Monthly Meteorological Magazine announced in 1871 that

A most violent storm of rain, hail and lightning visited Bath [England] on Saturday night. The rain descended in torrents. . . . The storm was accompanied by a similar phenomenon to that of the previous Sunday; myriads of small annelidae enclosed in patches of gelatinous substance, falling with the rain and covering the ground. These have been microscopically examined, and show, under a powerful lens, animals with barrel-formed bodies, the motion of the viscera in which is perfectly visible, with locust-shaped heads bearing long antennae, and with pectoral and caudal fin like feet. They are each an inch and a half long, and may be seen by the curious at Mr. R. Butler's, the Derby and Midland Tavern, where scientific men, on inspecting them, pronounce them to be marine insects, probably caught up into the clouds by a waterspout in the Bristol Channel.

According to an 1893 issue of *Nature* magazine, a thunderstorm on August 9, 1892, dropped living pond mussels on Paderborn, Germany: "A yellowish cloud attracted the attention of several people, both from its colour and the rapidity of its motion, when suddenly it burst, a torrential rain fell with a rattling sound, and immediately afterwards the pavement was found to be covered with hundreds of the mussels. . . . The only possible explanation seems to be that the water of a river in the neighbourhood was drawn up by a passing tornado, and afterwards deposited its living burden at the place in question."

Various newspapers and magazines tell of a rain of hazelnuts that fell on Dublin, Ireland, in 1867; a great quantity of grain that terrorized villagers in India when it fell on their town during a thunderstorm in March 1840; and a storm that deposited a covering of partially germinated Judas Tree seeds native to Central Africa on a town in Italy.

A widely repeated story of a rain of snakes that fell on Memphis, Tennessee, in January 1877 seems—thankfully—to be of doubtful authenticity. A report in the *Memphis Public Ledger* stated that "a decided sensation was produced in South Memphis by a heavy fall of small live snakes, thousands of which were to be seen on the ground this morning. . . . There are 'millions of them' on the ground yet, and the sight is not pleasant." The observations of other witnesses, however, suggest that the reptiles were not snakes, but common worm lizards, washed from a newly excavated road site by the heavy rains.

POSSIBLE PATHS
of HAILSTONES

wind

HAILSTONE CROSS SECTION

HARD RAIN: HAIL

Aristotle, the first of the great natural scientists, was baffled by hail. He had a relatively sound understanding of the mechanics of rain, which he said resulted from "vapour" rising into the sky, where it "cools and condenses again as a result of the loss of heat and the height and turns from air into water, and having become water falls again onto the earth." But hail resisted such easy explanation. In an age when Greek farmers tried to prevent hailstorms by burying laurel leaves soaked with menstrual blood in their fields, Aristotle was looking for rational answers. Hail was ice, obviously, yet it fell most often in warm weather. How does water in warm air become frozen? The only explanation, in Aristotle's view, was that warm air "compresses" cold air within a cloud, causing rain to freeze quickly into hailstones. He knew that tiny water droplets could ride aloft in a cloud in the same way grains of sand could float on the surface film of water, and that a number of droplets could coalesce into large raindrops. But "frozen drops cannot coalesce like liquid ones," he said. Thus, hailstones—even the largest of them—must be formed from raindrops that matched them in size.

Aristotle had no way of knowing it, but water droplets can become "supercooled" to temperatures far below freezing, and such droplets can indeed coalesce onto frozen drops. Updrafts within high, storm-bearing clouds are sometimes powerful enough to slow the descent of those drops, and perhaps to even toss them back upward, like popcorn in an air popper, passing them through alternating zones of cold and warm air and building up layers of ice, like onion skins, until the hailstones are heavy enough to fall to the ground.

Hail usually forms in warm weather, not because warm air refrigerates faster than cool air, as Aristotle thought, but because heat rising from the ground forms turbulent cumulonimbus clouds containing strong updrafts. The tops of those towering thunderheads can reach to the limits of the troposphere, where temperatures as low as −112 degrees Fahrenheit have been recorded.

In modern times the most widely accepted theory about hail formation states that ice pellets formed at the top of a cloud fall nearly to its base, picking up a layer of ice as they fall, then are caught in updrafts that carry them nearly to the top again to repeat the process. A more recent theory argues that a single hailstone makes only one descent from the top of the cloud to the ground, but that its fall is slowed by updrafts, allowing time—ten to twenty minutes by some estimates—for supercooled water droplets to adhere to the steadily growing nucleus of ice. Many meteorologists contend that both theories may be correct.

The powerful and unpredictable winds within a thunderstorm usually create hailstones that are roundish in shape, but they can also produce stones that are conical, disk-, star-, or pyramid-shaped, or irregular, with grotesque protuberances, and some have been reported perforated in the center, like Life Savers. Most of the hailstones that fall in North America have a diameter of less than one inch and most measure about one-half inch. They can be considerably larger, however. The largest ever measured under reliable circumstances fell on Coffeyville, Kansas, on September 3, 1970. It was preserved in a freezer, then sent to the National Center of Atmospheric Research in Boulder, Colorado, where it was found to weigh 1.67 pounds and measure 17.5 inches in circumference—about the size of a grapefruit.

Throughout history there have been numerous though usually unsubstantiated accounts of hailstones—or, more usually, aggregates of many hailstones frozen together—that make the Coffeyville specimen seem diminutive by comparison. One said to have fallen in Scotland was described in the *Edinburgh New Philosophical Journal* in 1849:

> Immediately after one of the loudest peals of thunder heard there, a large and irregular-shaped mass of ice, reckoned to be nearly 20 feet in circumference, and of a proportionate thickness,

fell near the farm-house. It had a beautiful crystalline appearance, being nearly all quite transparent, if we except a small portion of it which consisted of hailstones of uncommon size, fixed together. It was principally composed of small squares, diamond-shaped, of from 1 to 3 inches in size, all firmly congealed together. The weight of this large piece of ice could not be ascertained; but it is a most fortunate circumstance that it did not fall on Mr. Moffat's house, or it would have crushed it, and undoubtedly have caused the death of some of the inmates.

Similar huge masses of ice were reported falling in India in the eighteenth and nineteenth centuries. One was said to be the size of an elephant; another was reportedly twenty feet in diameter.

The biblical account of a hailstorm in the Book of Joshua is one of the earliest mentions of hail causing human deaths: "As they ran from Israel . . . the Eternal rained huge hailstones from heaven on them . . . they died of these. Indeed more died by the hailstones than at the hands of Israel by the sword."

Storms that produce hailstones as large as golfballs, tennis balls, and even baseballs are not uncommon. Less common, but not unheard of, are human deaths resulting from such formidable precipitation. Hailstones the size of goose eggs fell on the English army near Paris in 1360 and killed hundreds—some sources say thousands—of soldiers and horses, so disheartening the until-then victorious King Edward III that he agreed to sign the Treaty of Brétigny. British journalists in the nineteenth century reported that a hailstorm with stones measuring more than three inches in diameter killed 84 people and 3,000 oxen in the Himalayas on May 12, 1853. Hail the size of "cricket balls" reportedly fell on an area of six or seven square miles near New Delhi, India, on April 30, 1888, killing 246 people and more than 1,600 cattle, sheep, and goats. A hailstorm in Hunan Province, China, on June 19, 1932 killed more than 200 people and injured thousands. Hailstones the size of hen's eggs fell on Klausenberg, Rumania, on May 1, 1928, and killed six. A fall of large hailstones in Greece on June 13, 1930, killed twenty-two. And hailstones said to weigh from one to two pounds each killed twenty-three people in Rostov, Russia, on July

10, 1923. In the United States, the only known fatality from hail was a thirty-nine-year-old farmer who was killed near Lubbock, Texas, in May 1930 when he was caught in the open during a violent hailstorm.

Nature magazine reported in 1936 that a February 1 hailstorm that year in the Northern Transvaal region of South Africa killed nineteen people: "About three inches of rain fell in a few minutes, and then came the hail, which consisted of jagged lumps of ice. In thirty minutes the hail was lying everywhere to a depth of three feet, and in some cases the dead natives had to be dug out of it. There were many cattle killed, which the natives afterwards dragged away on sleighs. Whole crops were obliterated, and there are said to be over 1,000 families afflicted in the area."

Domestic and wild animals caught in the open sometimes suffer high mortality from hail. Biologists tracking the course of a hailstorm that battered a five-mile-wide path across Alberta, Canada, in July 1953, counted more than 36,000 dead ducks and ducklings. Just four days later another hailstorm struck the same region, killing at least 27,000 more waterfowl. In July 1978 some 200 sheep died when hail up to the size of baseballs pummelled parts of Montana.

In the United States, a 625-square-mile area known as Hail Alley, centered around the point where the borders of Nebraska, Colorado, and Wyoming meet, receives an average of nine or ten days of hail per year, more than any place else in North America. Farmers in that region are accustomed to seeing their wheat and corn crops battered by storms severe enough to leave hailstones piled on the ground like snowdrifts. A century ago, Italian grapegrowers who were confronted frequently with similar devastation of their vineyards, tried blasting storm clouds with cannons to prevent hail from forming. The cannons failed to stop the hail, but they undoubtedly made the farmers feel better.

STRONG *winds*

WEAKER *winds*   HORIZONTAL ROTATION

updraft

WALL
CLOUD

CYCLONE →

MESOCYCLONE

WARM AIR

NATURE'S TANTRUMS: TORNADOES

The earth, born in storm and upheaval, has always been battered by violent outbursts of wind and rain. Extremities of weather affect every inhabitant of the planet and influence every language. Because we know the frightening potential of that violence we can say with authority that we experience storms of passion, become embroiled in stormy relationships, fear the storm troops of invading armies. Our most common word for nature's periodic tantrums originated as *stirm*, an ancient northern European word which meant stirring foods together for cooking. Before evolving into "stir" and "storm," it was probably used as a vivid metaphor, as in "the stirring of the skies."

Few storms stir the skies with the power and violence of a tornado. In photographs a sinuous funnel cloud sweeping across the prairie is captivating and beautiful. But witnesses who have seen the same cloud up close and first-hand always repeat the same message: Tornadoes are terrifying. Afterwards they tend to describe what they have seen in anthropomorphic terms calling the tornado a "beast" or "monster" that emitted sounds like the roar of a thousand freight trains, a B-52, a dog's howling, a million bees, the clattering of thousands of Venetian blinds, a dozen factories filled with buzzsaws, or the booming of a thousand cannons. Most people who have encountered a tornado never want to see another.

Still, there is no denying that tornadoes are among the most fascinating of weather anomalies. An average of 850 twisters touch down in the United States each year, yet they are uncommon enough to be newsworthy wherever they appear. Their tightly wound vortices, usually measuring less than a few hundred feet across, are enigmatic and unpredictable performers of strange and destructive feats.

My father was a police officer in Flint, Michigan, when a tornado destroyed part of that city in 1953. His partner happened to be off-duty that day at his home in a neighborhood that lay in the path of the tornado. He would later tell my father that when the day grew dark and ominously still he stepped outside into his yard and noticed the sky had turned a strange, sickly green. Then he heard what sounded like a freight train bearing down on him and before he could run inside, houses two or three doors away exploded, and he was suddenly lifted and pinned against the wall of his house, as if he was being held off his feet by a bully. He recalled that he watched a tree in his yard bend and twist so violently that the individual fibers of wood in the trunk separated the way the strands in a rope separate if you twist it. When the wind released the tree it returned to its original shape, with no sign that it had been split apart, except that from the trunk emerged pieces of grass and stalks of weeds that had been caught there when the tree closed up. They looked for all the world as if they had been driven in place by the wind.

That is not to say that the stories of straws driven into telephone poles and stalks of wheat piercing tree trunks are not true. Pieces of lumber clocked at 100 miles per hour can blast through concrete

blocks, and winds in a tornado are certainly stronger than that. With winds so powerful and capricious as a tornado's, strange things are certain to happen. People caught in them are sometimes tattooed for life by sand and grit driven deeply beneath their skin. Tornadoes have been credited with plucking the feathers from chickens, shearing the wool off sheep, driving wooden planks through steel girders, sucking soda from open bottles, lifting sleeping infants from their cradles and depositing them unhurt—and still asleep—hundreds of feet away. They have vacuumed buckets of water from open wells, lifted dressers in the air and set them down a hundred yards away without cracking their mirrors, and pulled fence posts from the ground. Where and when they will strike is almost impossible to predict. In the kind of bizarre coincidence that could make even the staunchest rationalist knock on wood and rub a rabbit's foot, the town of Codell, Kansas, was struck by twisters in 1916, 1917, and yet again in 1918—each year on the same day, May 20.

A number of complex and not entirely understood conditions must be met before a tornado can form. The first necessity is a mesocyclone: a column of rising warm air that rotates because of conflicts in the speed and direction of winds at various altitudes. The rotation may begin as a horizontal spin, like a tube rolling along the ground, caused by strong winds blowing over weaker ones. If the tube of rolling air is picked up by a powerful updraft the horizontal rotation can convert to vertical rotation strong enough to start the entire storm rotating. As it rotates, additional warm air is sucked into the rising column of air with enough force to create updrafts up to 100 miles per hour for thousands of feet to the top of the thunderstorm.

Once a tornado is born, the spinning air causes atmospheric pressure to drop in the center of the funnel. As the pressure drops the air expands and cools, causing the moisture within it to condense into a visible funnel cloud. A "pure" tornado is often white, but it will turn gray, black, or brown as it picks up dirt and debris from the ground and as the clouds above it are drawn downward.

One peculiarity of mesocyclones that has long baffled meteorologists is that only about half of them produce tornadoes. Scientists desperate to discover a triggering influence once proposed that the tiny eddies left in the wake of speeding automobiles might be the missing catalyst in tornado production. Others have argued more convincingly that the contact between warm air rising and cool air falling causes winds between them to spin in the same way a top can be made to spin by rolling it between your hands.

Tornadoes usually appear at the rear of a mesocyclone, trailing behind the darkest, fiercest clouds and the heaviest falls of rain and hail, and usually descend first in a slowly rotating appendage known as a wall cloud. One clue to a tornado's formation might be that a downdraft of cool air is often found at the rear of the storm system, falling downward and being spun into the warm inflow at the base of the storm. In a typical thunderstorm—and only about one thunderstorm in every 1,000 produces a tornado —the downdraft and updraft mingle, eventually causing the storm to collapse. In the severe thunderstorms known as *supercells*, however, where the most severe tornadoes are born, the updraft and downdraft are separated, and capable of feeding one another for hours at a time.

Tornado researcher T. Theodore Fujita of the University of Chicago discovered in the 1970s that the largest and most destructive tornadoes often contain as many as five or six smaller twisters within the main funnel. He called the smaller tornadoes *suction vortices*, and determined that they usually measure less than thirty feet in diameter and are formed by a strong downdraft at the center of the primary tornado that meets the warm air rushing upward. It is such miniature tornadoes spinning around inside, at the edges, or straying outside the main vortex that cause much of the erratic damage so often reported from tornadoes, such as one house being completely destroyed while another across the street loses not even a shingle from its roof.

Until recently, it was speculated that winds within a tornado could reach speeds of 500 or more miles per hour. Researchers found, however, than even winds of 100 miles per hour are capable of inflicting enormous damage, and that the strongest tornadoes probably never generate ground winds in excess of 250 or 260 miles per hour.

The Coriolis effect of the earth's rotation causes mesocyclones north of the equator to spin counterclockwise, and those south of the equator to spin in a clockwise direction. Tornadoes tend to duplicate the direction of their parent storm, so that the majority of twisters in North America spin in a counterclockwise direction.

The life span of a tornado is usually brief, seldom more than a few hours, and usually covering only five to fifteen miles of ground. The mesocyclone weakens and collapses, or the bottom of the tornado is drawn away from the top, slowed by friction on the ground or sped away by ground winds. The funnel cloud that had descended more or less vertically from the cloud wall, grows to a diagonal or even nearly horizontal angle to it. The final "rope" stage causes the funnel to lose power as it shrivels, narrows, and becomes elongated.

The United States experiences more tornadoes than any other country, but they occur at least occasionally in many other parts of the world, including Russia and the Baltic States, England, Australia, India, China, Japan (where they are sometimes called *dragonwhirls*), Bermuda, the Fiji Islands, and South Africa. In the United States, they have been reported in every state, in every month, at every hour of the day. One was once seen sweeping across Utah's Wasatch Mountains in winter, sucking up snow as it went, yet most of the 600 to 1,000 tornadoes that touch ground each year in the United States occur in afternoons and early evenings in May and June. In the Gulf Coast states they are most frequent in late winter and early spring, when warm, moist air from the Gulf of Mexico sweeps northward and collides with cool, dry air from Canada. In late spring and into summer, as the jet stream moves north, severe thunderstorms and tornadoes begin showing up with increasing frequency in the central plains, especially in a strip of land from Texas to Nebraska known as Tornado Alley, where about one-third of U.S. tornadoes occur each year. Later yet, from midsummer into autumn, they appear more frequently in northern regions around the Great Lakes and in the Dakotas.

Because the conditions that make tornadoes possible often occur across a wide front, tornadoes sometimes appear in clusters. On April 3 and 4, 1974, a total of 148 twisters struck from Alabama north through Tennessee, Kentucky, Illinois, Indiana, Ohio, and Michigan, killing 315 people and destroying thousands of homes. It was the largest outbreak of tornadoes ever recorded in a single twenty-four-hour period.

The most devastating tornado on record ripped a trail a mile wide and 219 miles long through Missouri, Illinois, and Indiana on March 18, 1925, killing 689 and injuring nearly 2,000.

Most of us probably visualize the inside of a tornado as something like what Dorothy saw in *The Wizard of Oz*, but a more reliable eyewitness account was recorded by a Kansas farmer named Will Keller in 1928. A tornado bore down on his farm on June 22 of that year, and he and his family took cover in their storm cellar. As he was about to close the door of the cellar the tornado lifted off the ground and passed over him. Later, he described what he saw:

Steadily the tornado came on, the end gradually rising above the ground. I could have stood there only a few seconds but so impressed was I with what was going on that it seemed a long time. At last the great shaggy end of the funnel hung directly overhead. Everything was as still as death. There was a strong gassy odor and it seemed that I could not breathe. There was a screaming, hissing sound coming directly from the end of the funnel. I looked up, and to my astonishment I saw right into the heart of the tornado. There was a circular opening in the center of the funnel, about fifty to one hundred feet in diameter and extending straight upward for a distance of at least half a mile, as best I could judge under the circumstances. The walls of this opening were rotating clouds and the whole was brilliantly lighted with constant flashes of lightning, which zig-zagged from side to side. Had it not been for the lightning, I could not have seen the opening, or any distance into it. Around the rim of the great vortex small tornadoes were constantly forming and breaking away. These looked like tails as they writhed their way around the funnel. It was these that made the hissing sound. I noticed the rotation of the great whirl was anti-clockwise, but some of the small twisters rotated clockwise. The opening was entirely hollow, except for something I could not exactly make out but suppose it was a detached wind cloud. This thing kept moving up and down. The tornado was not traveling at a great speed. I had plenty of time to get a good view of the whole thing, inside and out.

Waterspouts and Wind Devils

VORTEX MOTIONS, from hurricanes to tornadoes to swirls of dust in a dry parking lot, are a common feature of the atmosphere. Far less destructive and more frequently seen than tornadoes are their smaller cousins, waterspouts and wind devils. Pliny described waterspouts as water drawn up by clouds "like a pipe," and included them with whirlwinds and cyclones as a maritime phenomenon that is "specially disastrous to navigators, as it twists round and shatters not only the yards, but the vessels themselves, leaving only the slender remedy of pouring out vinegar in advance of its approach, vinegar being a very cold substance. The same whirlwind when beaten back by its very impact snatches things up and carries them back with it to the sky, sucking them high aloft."

Waterspouts occur with some frequency over most of the oceans of the world, as well as over inland lakes and large rivers. I once saw one on Lake Michigan as I stood on shore two or three miles away. Trailing behind a rain squall like a reluctant dog on a leash, it extended in a thin, nearly transparent column perhaps 500 feet from the cloud to the water. Where it touched the surface of the lake it raised a mushroom of spray about 100 feet across, which was picked up and carried in a loose translucent spiral toward the clouds. As the spout approached land it gradually lifted off the lake, losing its momentum, and the water it had previously raised fell as rain on the beach.

Although ordinarily considered much less dangerous than tornadoes, waterspouts do their share of damage. Five ships were reportedly sunk by a waterspout in the harbor at Tunis in North

Africa in 1885, and a ship at sea was struck by a waterspout 500 feet in diameter that stripped away the masts and sails and blew a crew member overboard.

Some waterspouts are essentially no different than tornadoes. One that occurred near St. Petersburg, Florida, on June 13, 1952, described in *Weatherwise* magazine, seems to fit the category: "it made a terrific roar as it passed over the shallow water [of Big Bayou]. A wave five feet high was rushing ahead of it . . . as it hit land the earth appeared to explode. Trees, sand and all manner of material roared skyward. It disappeared in a mist a few minutes later."

Others, however, seem to be marine phenomena. Many occur in midsummer, in regions of high water temperature near the equator, and sometimes when there are no clouds in the sky. They can be composed of gentle winds barely capable of lifting spray from the surface, or of a powerful vortex that lifts saltwater high into the air and can raise a column of solid water ten or twenty feet in the center of the vortex.

Whirlwinds and dust devils are common in desert regions, especially during storms, when powerful winds form eddies as they blow around obstructions, or when intense heat from the sun creates small pockets of atmospheric instability. Dust devils are the smallest of the world's rotating, convection-fed storms and can take a variety of forms. Some are hollow, with a spiral of dust rotating on the outer edge only; others are virtually filled with dust and other debris. Most dust devils are short-lived—usually lasting only a few minutes—and relatively harmless, but they occasionally grow to a height and power sufficient to lift large quantities of soil and dust, strip shingles from roofs, and uproot small trees.

Birds

In spring, when love is in the air, it is literally in the air much less often than you might expect. Birds involved in courtship and mating are at their most vulnerable and few make a public, airy display of their sexual behavior. Most feathered sex takes place on the ground or among trees, where predators are least likely to be attracted. But a few birds—and more than a few insects—are so enamored with flight that they have not separated it from the lively business of reproduction.

The mating habits of birds are understandably difficult to observe, and there is a lack of definitive information on the subject. Some of the blame for misinformation must go to that enthusiastic but woefully careless chronicler of nature, Pliny the Elder. Consider this breathless passage about the sexual proclivities of partridges, from his *Natural History*:

> The cocks, owing to their desire for the hens, fight duels with each other; it is said that the one who loses has to accept the advances of the victor. . . . And in no other creature is concupiscence so active. If the hens stand facing the cocks they become pregnant from the afflatus that passes out from them, while if they open their beaks and put out their tongue at that time they are sexually excited. Even the draught of air from cocks flying over them, and often merely the sound of a cock crowing, makes them conceive. And even their affection for their brood is so conquered by desire that when a hen is quietly sitting on her eggs in hiding, if she becomes aware of a fowler's decoy hen approaching her cock she chirps him back to her and recalls him

MAYFLY LIFE CYCLE

AIRBORNE DRAGONFLIES

and voluntarily offers herself to his desire. Indeed they are subject to such madness that often with a blind swoop they perch on the fowler's head.

Ornithologists have come a long way since the days of Pliny in their understanding of bird behavior, but even now there are many unanswered questions. As far as anyone knows, swifts and perhaps some swallows are the only birds that throw prudence to the wind and actually mate in midair. The swifts are an aptly named family of long-winged, fast-flying, swallowlike birds found in many parts of the world. Most of the numerous species of swifts feed by "hawking" for flying insects, and are among the fastest and most adept flyers in the world. The Eurasian swift, which is found in most of Europe and parts of North Africa and Asia, virtually lives in the air, doing all its feeding, drinking, sleeping, and mating on the wing; it lands only to lay its eggs and feed its young. Pairs of the white-throated swift of the western United States and Mexico have been observed during early mornings in the spring swooping down canyons at estimated speeds of forty miles per hour, meeting briefly and tumbling through the air in apparent copulation. The mating involves little more than a brief contact of both birds' swollen sex organs, or cloacals, a moment of intimacy described poetically by ornithologists as a "cloacal kiss."

Among birds, aerial courtship displays are far more common than aerial copulation. Courtship rituals serve to attract birds of the same species—and to encourage mating of the healthiest, most robust individuals. Males of many species use such displays to attract mates as well as to chase off rivals and establish territories. Perhaps not surprisingly, those birds that are the most skillful flyers usually indulge in the most elaborate aerial displays. Bank swallows engage in complex flights that include passing a white feather back and forth in flight. Many species of male hummingbirds attract the attention of females by hovering before them, swinging pendulum-like, then rising in swooping, arching, or figure-eight flights from a few feet to as much as 100 feet above them. If a female is attracted, she might join the male in his prenuptial dance before they drop to the ground and copulate.

The sky-dance of the American woodcock is an unusual court-ship ritual that is relatively easy to observe in some parts of the northeastern and northcentral United States. It takes place during the evening and around dawn in April and May and is often re-peated year after year in the same places. Renowned conservation-ist Aldo Leopold described it with typical enthusiasm in *A Sand County Almanac:*

> He flies in low from some neighboring thicket, alights on the bare moss, and at once begins the overture: a series of queer throaty *peents* spaced about two seconds apart, and sounding much like the summer call of the nighthawk.
>
> Suddenly the peenting ceases and the bird flutters skyward in a series of wide spirals, emitting a musical twitter. Up and up he goes, the spirals steeper and smaller, the twittering louder and louder, until the performer is only a speck in the sky. Then, without warning, he tumbles like a crippled plane, giving voice in a soft liquid warble that a March bluebird might envy. At a few feet from the ground he levels off and returns to his peenting ground, usually to the exact spot where the performance began, and there resumes his peenting.

Aerial courtship can sometimes be observed among ospreys, kestrels, falcons, kites, and nighthawks, all of which attract mates on the wing but copulate on the ground or in trees. The male red-shouldered hawk soars and swoops, climbs to heights up to 2,000 feet, then tucks its wings and dives nearly to the ground. The male northern harrier's courting flight has been described as a series of capital Us strung together: UUUUUUUUUUU, with the bird rising to about 500 feet, stalling at the apex, sometimes making a somer-sault, and plummeting toward the ground. Males of some species of owls soar and swoop, and often clap their wings together below their bodies.

The most awe-inspiring aerial courtships of all might be the rarely witnessed "whirling display" of the bald eagle, in which the male and female lock talons and descend hundreds of feet together in a series of somersaults. Myrtle Jeanne Broley described such a scene in her book *Eagle Man: Charles L. Broley's Field Adventures with American Eagles*: "Just then he heard an eagle scream and, looking up, saw one of the two he had observed before chasing the other. Suddenly they came together, locking talons and executing four complete cartwheels, tumbling down several hundred feet before breaking apart. . . . He has been lucky enough to see this courtship manoeuvre three times. Doctor Arthur Allen of Cornell has seen it once, and a number of other naturalists have told us about seeing it."

In the nineteenth century, Walt Whitman described the same whirling courtship in his poem "The Dalliance of the Eagles," which he based on an eyewitness account by his friend, the naturalist John Burroughs:

Skirting the river road, (my forenoon walk, my rest,)
Skyward in air a sudden muffled sound, the dalliance of the eagles,
The rushing amorous contact high in space together,
The clinching interlocking claws, a living, fierce, gyrating wheel,
Four beating wings, two beaks, a swirling mass tight grappling,
In tumbling turning clustering loops, straight downward falling,
Till o'er the river pois'd, the twain yet one, a moment's lull,
A motionless still balance in the air, then parting, talons loosing,
Upward again on slow-firm pinions slanting, their separate diverse flight,
She hers, he his, pursuing.

Insects

LIKE FREE-FLYING BIRDS that seem to delight in showing off their flight skills to prospective mates, monarch butterflies, among the most graceful and tireless of all the lepidoptera, put on entertaining courtship displays during the summer. Look for them among blossoming flowers in meadows or along roadsides. The males often perch on the tips of branches or the tops of weeds, waiting for other butterflies to sail past. When one comes near, the male flies out to see if it is a female of his own species. If it is, he pursues her, bumping the tip of her abdomen. They then chase each other in an erratic and rapid flight through meadows, over bushes, through tree branches, and upward to heights of a hundred feet or more. Finally, the male grasps the female from above, while holding his wings stationary and outstretched. The female flutters her wings enough to keep the two of them upright, and together they glide to the ground. Copulation commences there, but the male often launches them into flight again, this time carrying the weight of the female while she leaves her wings closed and legs folded beneath her.

Insects are far more likely to be seen mating in flight than birds. Many species of dragonflies, for instance, copulate on the wing, and can often be heard before they are seen, the combined wing-beats of a conjugal pair creating a startling whirring buzz. The 4,700 species of dragonflies found worldwide are descendants of prehistoric species at least 250 million years old that included what were probably the largest insects that ever lived. The largest fossilized remains belong to the giant dragonfly *Meganeura*, which had a wing span of nearly 2½ feet. At various times dragonflies have been known by a remarkable number of names. A natural history text published in 1603 lists them as "parsons, wild horses, camels, water sprites, naiads, and water vicars." In more recent times, especially in North America, they have been called mosquito hawks, snake doctors, horse stingers, sewing needles, and Devil's darning-needles, in reference to the distinctive shape of their elongated abdomens, and probably also to the age-old and entirely mistaken belief that they are capable of inflicting stings.

A dragonfly begins its life as a ferocious predatory larva, known as a niad, that searches the water for prey by expelling water through gills at the end of its abdomen, propelling it, in the words of J. Henri Fabre, like a "hydraulic cannon." The niad captures insect larvae, tadpoles, and small minnows with a special prehensile mask that it can extend rapidly to pierce and hold prey. After one or more years of aquatic life, the larva climbs out of the water onto a stem of a plant or onto the land, splits its skin, and metamorphoses into an adult dragonfly.

Dragonflies mate by coupling in midair for periods ranging from five seconds to a half hour or more. The act was described color-fully and accurately by the seventeenth century Dutch naturalist Jan Swammerdam:

> The male, as it whirls in the air in a series of rapid turns, is able to extend its tail to the female with dexterity beyond all measure. The latter seizes it to the point between her head and eyes, push-ing it to the back of her neck, and grasps it with her legs very eagerly and avidly. Once she has a firm hold on the tail, she curves her abdomen forward to the male copulatory structures that lie at the front on its breast. Consequently, union occurs during flight as they swarm in the air. The extremity of the fe-male's tail is curved up against the mid-part of the male body where the latter conceals his penis. This penis penetrates the genitalia of the female that are situated at the end of the female tail. And in order that the female can reach as far as the breast of the male [where the penis is located], she contracts her body and curls her tail in a marked curve. After fertilization has been achieved in this way, the female finally dips her tail into water and rapidly deposits her eggs there.

Midair mating slows flying insects and makes them conspicuous, thus making them especially vulnerable to birds and other preda-tors. Dragonflies are such rapid, skilled flyers that they can afford the risk. Other insects, like the mayflies, can succeed only because they mate in swarms containing so many individuals that an ade-quate percentage are likely to survive long enough to lay their eggs. Mayflies are of the order *Ephemeroptera*, from the Greek *ephemeron*, "living only for a day," in reference to the brief lifespan—from about an hour to a few days—of the mature adult, or imago. Aris-totle was fascinated by these short-lived and elegant insects, and described them in detail. Later naturalists, as often happened, based their own discussions of the insects more on a blend of watered-down Aristotle, hearsay, and folk tales than personal ob-servation. The Roman writer Claudius Aelianus, 600 years after Aristotle, would describe an insect only barely resembling those Aristotle had seen hatching from the river Hypanis: "There are creatures called Ephemera that take their name from their span of life, for they are generated in wine, and when the vessel is opened they fly out, see the light, and die. Thus it is that Nature has permitted them to come to life, but has rescued them as soon as possible from life's evils, so that they are neither aware of their own misfortune nor are spectators of the misfortune of others."

The mayflies are a large order of about 2,000 species found worldwide. They spend most of their lives—one to three years—as nymphs in streams, ponds, or lakes where they burrow into mud or sand, or cling to rocks and vegetation, breathing through gills and feeding on algae, microorganisms, and decaying plant and animal material, while molting as many as twenty times as they grow. In most species the nymph rises to the surface in the spring (though some species emerge in summer, fall, and even winter), and molts into a winged subimago—a subadult stage that is unique among winged insects. The subimago flies to shore and takes cover in brush or trees for a few hours or a few days, then undergoes another molting and emerges as a sexually mature imago.

Mating takes place near shore over water, starting with a swarm, or "mating aggregation," of males numbering up to the millions. They hover suspended in the air or move rhythmically up and down, each male climbing to a height of ten or twenty feet, then allow themselves to fall a few feet with wings and tails raised. Mayflies at this stage have no interest in anything but mating. They have no mouthparts for feeding and their digestive system contains only air to help make them lightweight and agile in flight. The males have keen vision, thanks to extremely developed compound eyes, and are able to quickly spot females. When a female is attracted to the swarm, or blunders into it, she is quickly seized from below by a male, and held at the head with his long forelegs. The male presses his back against the female's abdomen—or, in a few species, presses belly to belly—and copulation occurs during a few seconds in flight, before the mating pair loses height and separates. Within minutes the female oviposits her eggs, ejecting them over water or dipping her tail in the water to dislodge them or, in a few species, crawling into the water to deposit them. Within a few minutes the male and female collapse on the water "spent," their wings outstretched, and die.

Mayflies can be among the most abundant insects in a body of water, and their mating swarms can be so immense they create traffic hazards. At night, mating mayflies are attracted to light and sometimes mistake gleaming highways for rivers, resulting in mass die-offs that can cause roads to become dangerously slippery. In some places so many dead mayflies have accumulated that snow shovels and plows have been used to clear clogged streets.

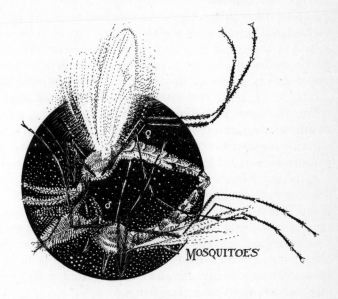

MOSQUITOES

Even the largest swarms of mayflies are no cause for alarm, but one insect that mates on the wing is responsible for at least a million human deaths every year. If ever a swarm of insects should strike fear in your heart, it is a swarm of mating mosquitoes. These tiny, seemingly innocuous pests are famous for raising irritating and itching welts on the skin of the warm-blooded hosts they feed on. The name of one of the more than 3,000 species worldwide, *Aedes vexans*, translates to "disagreeable annoyance." But mosquitoes are more than a mere annoyance. In their business of inserting a hypodermiclike mouthpart known as a proboscis into the bloodstream of a host, they can transmit a witch's brew of deadly illnesses, including malaria, yellow fever, dengue fever, encephalitis, filariasis, and sleeping sickness. The ancient Greek empire may have been crumbled by malarial mosquitoes. Yellow fever transmitted by mosquitoes killed more than 20,000 during the building of the Panama Canal. Mosquitoes were responsible for more casualties in the Pacific Theater during World War II than all the guns and bombs combined.

Among mosquitoes, only the females feed on blood. Males sip the nectar of flowers. Both sexes begin their lives as larvae in lakes, ponds, or any still or stagnant body of water—even tiny puddles in the crooks of trees or the insides of discarded tin cans and automobile tires—where they hang suspended near the surface and feed on bacteria, protozoans, microplankton, and dead vegetation. The larvae molt three times as they grow, then molt into pupae, which, after a few days, molt yet again into adults. Mating is initiated by swarms of males gathering over water in the evening. Females are attracted to the swarm, enter, and in turn attract mates with their high-frequency buzzing.

In spite of enormous efforts to eradicate mosquitoes, there seems little hope of doing away with them for good. They have the uncanny ability to mutate into forms that resist the effects of insecticides and to breed in the least likely of places. Even if our planet should become inhospitable to most forms of life, several species of mosquito have shown a preference for breeding in water that is unsuitable to other living things.

SUMMER

23½°

Summer · Solstice

INTRODUCTION

In summer the days seem to stretch out forever and living, as the song says, can be easy. It is a season of abundant life and languid and sensual revelry. Yet those days are tainted with the knowledge that they will soon be growing shorter and the nights longer, an ambivalence that may be at the root of ancient beliefs that during the hottest part of summer dragons patrolled the sky and witches were active. Midsummer Eve was celebrated on June 23, the eve of the nativity of Saint John the Baptist, and was said to be the witches' sabbath, when evil could fall upon the season's crops in the critical few weeks before harvest. To protect the crops and throw some sympathetic magic toward the diminishing sun, bonfires were lit and the smoke from torches was allowed to drift downwind over the fields.

A girl wanting to conjure some beneficial magic on Midsummer's Eve could gather yarrow from a young man's grave and put it beneath her pillow to give her dreams of her husband-to-be. Sprinkling yourself with fern seed on Midsummer's Eve could make you invisible. If you walked backward that night and cut a branch of hazel from between your knees it would lead you to hidden treasure. To test the branch's potency hold it near water: If it squeaks like a pig it is highly charged stuff.

The summer solstice occurs in the Northern Hemisphere on June 20 or 21, and marks the official beginning of summer in most countries. It ends about September 23, the autumnal equinox, but the season is highly variable and commences and ends at different times throughout the world. In North America, if you think of summer as the warmest three months of the year, it begins about May 25 in Texas, Arizona, and New Mexico; about May 27 in Florida, Georgia, Louisiana, and Nevada; about June 7 in North Dakota, Minnesota, Wisconsin, Michigan, New York, and New England;

about June 15 in northern Maine and the Maritime Provinces of Canada; and about June 21, the day of the solstice, in subarctic Quebec, Labrador, and Newfoundland.

The word *solstice* is derived from the Latin *solstitium*, which translates as "sun standing still." Anyone willing to observe the rising or setting sun every day for a few weeks before the summer (or winter) solstice can watch the sun "standing still," and understand better the source of the Latin word and the many references to the phenomenon in world mythology. Choose a spot to stand every day—a hilltop or a doorstep, it doesn't matter, as long as you have a good view of the sun on the horizon—and note the position on the horizon where the sun rises or sets. It helps to draw a silhouette of the horizon, and mark the sun's location each day, judging its movement against landmarks such as hills, buildings, and trees. If you watch every day for a year, you will see that the sun travels up and down a length of horizon, with a northern and southern boundary. The farther north you live, the greater the distance the sun travels each year, whereas at the equator it rises and sets at nearly the same position every day, year-round. At the March and September equinoxes, the sun's passage is the most noticeable of the year, appearing farther north every day in March (thus increasing the amount of sunlight every day) and appearing farther south every day in September. But as the summer solstice approaches, the rate of difference slows, in much the way a pendulum slows as it reaches the limits of its arc. By the solstice, the sun appears to stop for three or four days, rising and setting at nearly identical positions on the horizon—just as a pendulum stops at the end of its swing or a ball thrown straight up in the air stops as it reaches the peak of its flight. Once the sun has reached its peak at the solstice, the situation reverses, and it moves slightly farther to the south every day, resulting in increasingly shorter hours of sunlight.

It is logical to conclude that the June solstice, which many cultures celebrate as the first day of summer, is in fact the *middle* of summer, since it is the point at which the sun is highest and remains longest in the sky. Nonetheless, in northern regions we find it natural to think of the solstice as the beginning of the season because the hot weather of summer usually follows a few weeks afterwards (the lag is caused by the insulating effect of the surface of the earth, which holds winter's cold even after the sun's warmth has increased). In southern temperate regions, June 21 is more likely to seem the height of summer.

The day of the solstice begins slightly earlier than the day that precedes it, and ends slightly later. It is a daylight-lover's delight, the longest day and shortest night of the year. But celebrations of the summer solstice are tempered with a bittersweet irony: The day following the solstice begins a bit later and ends a bit sooner, and the march toward winter has begun.

TAKING THE HEAT

Say summer and most people think of heat. In the temperate regions, this season finds the earth tilted toward the sun in such a way as to increase both the duration and intensity of sunlight, resulting in the hottest temperatures of the year. How hot? The world's record was set on September 13, 1922, in El Azizia, Libya, where it reached 136.4 degrees Fahrenheit (in the shade). The hottest spot in the Western Hemisphere is Death Valley, California, where on July 10, 1913, the shaded temperature reached 134 degrees; during one six-week period in Death Valley the temperature rose to at least 120 degrees every day. Marble Bar, Australia, is on record with the longest hot spell: In 1923, it suffered through 162 consecutive days of 100-degree or higher temperatures.

We have inherited the term "dog days" from the ancient Romans, who believed that the high temperatures common from about July 3 to August 11 were the result of stellar heat given off by Sirius, the Dog Star, the summer sky's brightest star. In fact, of course, all heat has a solar source, though by late summer we get a double dose, both direct heat from the sun and radiant heat rising up from the earth. Soil and water are efficient reservoirs of heat. After weeks of baking under the summer sun the ground is saturated with so much heat that it contributes to sleepless nights by giving it back off to the atmosphere.

Anyone who is even moderately active on hot summer days knows what it is like to perspire buckets, to be washed in rivers of sweat. Mow the lawn or play a set of tennis and you end up soused, soaked, stewed, steeped, dunked, and drenched in it.

The object of all this fluidity is temperature regulation, a critical faculty among animals like us, with our relatively narrow tolerance of temperatures. Cool our bodies more than 15 to 20 degrees below 98.6 and we lose consciousness and die; heat us more than six or eight degrees above normal and we suffer the dangerous symptoms of heat stroke. We are hardy creatures, but dreadfully intolerant of extremes. Perspiration cools us because evaporating water unleashes heat from the skin and from the blood flowing through capillaries within it. Wear tight-fitting clothes and the benefits are mostly lost. The clothing becomes soaked with perspiration, but because circulating air cannot reach the wet skin, we are likely to become dehydrated without the benefit of cooling. Which is why desert-dwelling people have for thousands of years worn loose, flowing garments that allow air to pass over the skin. Strangely, when those loose garments are dark-colored their cooling effect is enhanced. The desert tribes of the Sahara discovered centuries ago that black robes and tents can actually be cooler than light-colored, reflective fabrics—as long as there is a breeze. Solar heat is concentrated near the surface of black objects, and is quickly carried away by wind, reducing further penetration of radiant heat. The black desert raven can therefore be cooler than light-colored birds.

Sweating is an effective way to cool the body, but is wasteful of water, an important consideration in those parts of the world where heat's diabolical half-sister is drought. Only large mammals sweat, presumably because only creatures with a relatively large ratio of body mass to surface area can afford the luxury. Smaller creatures rely on less wasteful measures. Dogs pant—and not just setters and dachshunds, but all the canids, from African bush dogs to arctic wolves. Panting transfers excess heat from the lungs and tongue, and performs essentially the same function perspiration does in larger mammals.

Birds, too, use panting as their chief means of cooling off. Because most desert birds are active in daylight (owls and nightjars are the exception; they take cover during the day in burrows in cacti or in clefts and fissures in rocks) and are therefore exposed to a great deal of radiant energy from the sun, they are limited to living in areas with consistent water supplies. Even where water is abundant birds are forced to cool themselves by making adjustments to allow air movement or reduce sunlight. Sea birds in the tropics elevate their plumage to expose skin to cooling breezes. Nesting gulls rotate constantly to remain facing the sun, offering the smallest amount of body area to direct sunlight. Cormorants, frigate birds, and boobies are especially vigilant at protecting their nestlings from heat by shading them with outstretched wings. Black vultures excrete on their legs to increase the body heat lost through evaporation. The Galapagos penguin, among the smallest of penguins, has adapted to the heat of the equatorial Galapagos Islands by becoming mostly nocturnal and nesting in cool burrows.

The Namaqua sand grouse of the Kalahari Desert beats the desert heat during the nesting season by supplying young nestlings with an ingenious source of water. Each day the male of a nesting pair flies as far as fifty miles to the nearest water hole, where he soaks his belly feathers in water, then flies back to the nest and allows the nestlings to suck the water from the feathers. Although more than half the moisture evaporates before the adult male reaches the nest, his offspring manage to slake their thirst with about half an ounce of water each trip.

The best way to beat the heat is to avoid it. Desert animals tend to be nocturnal, passing the daylight hours burrowed safely out of reach of the sun's heat. The fennec fox of the Sahara spends its days underground and during the night feeds on insects, lizards, rodents, and plants, acquiring from this diet virtually all the water it needs. Ground squirrels are diurnal, but stay in burrows during the hottest part of the day and are fully active only in early morning and late afternoon. Desert hares do not burrow to escape midday heat, but take cover in shaded depressions where their enormous ears help radiate heat away from the body. Grant's gazelle, an inhabitant of semiarid northern Kenya, ingests most of the water it needs by feeding at night. During the day, the plants the gazelle favors contains only about 3 percent water and are so dry they crumble when touched. After eight hours exposure to night air, however, the same plants contain 40 percent of their weight in water, and provide the gazelle with most of the three liters of water it needs each day.

Rodents are among the most successful desert animals, spending their days burrowed underground to avoid high temperatures and low humidity. The North African sand rat survives on a diet of succulent, saline plants—eliminating the excess salt in highly concentrated urine. The American grasshopper mouse lives almost entirely on insects, from which it gets most of the moisture it needs.

The champion of the desert rodents must certainly be the kangaroo rat of the American Southwest, with its remarkable ability to live virtually without external sources of water. During the day, a kangaroo rat retreats to its burrow, plugging the entrance to keep heat out and humidity in. As long as it remains in the relatively cool and damp burrow during the hottest hours of the day it can produce all the water it needs by oxidizing it from a diet of dry grains and seeds. Oxidization produces metabolic water by combining oxygen captured during respiration with hydrogen found within the dry food. The rat manages the delicate balance necessary for survival only by conserving water in every possible way. Its urine is extremely concentrated; its feces almost dry. High humidity in the burrows—two to five times the moisture content of outside air—reduces the rate of evaporation from the lungs. By staying underground during the day, the rat produces more metabolic water than it uses, resulting in a net gain.

Invertebrates must work harder to prevent water loss by evaporation than larger creatures, since the surface area of their bodies is greater in relation to their mass. In arid regions, spiders, scorpions, and many insects, most of them nocturnal, secrete a thin layer of wax that is relatively impermeable to water vapor. Because the wax also prevents the transfer of oxygen and carbon dioxide, the invertebrates have evolved a respiratory mechanism that works like a valve, opening periodically to allow carbon dioxide out and oxygen in. Insects and arachnids also excrete dry fecal matter, or may store the excrement in tiny pellets within their bodies. They eliminate the waste when water is abundant or, in some cases, never eliminate it at all.

In summer the living is easy, sure, but not always comfortable. Go out in the heat of summer's dog days and temperatures rise, tempers flare, riots erupt. Better to make like a kangaroo rat and burrow. The Greek poet Hesiod, a sensible gentleman indeed, had this advice for getting through summer's worst: "When Sirius parches head and knees, and the body is dried up by reason of the heat, then sit in the shade and drink."

Too Much Heat:
The Dangers of Global Warming

FOR THE ATMOSPHERE to continue working the way we are accustomed to seeing it work, it must maintain a fairly consistent chemistry. A large or sudden increase or decrease of any one ingredient can throw the entire system out of balance, with frightening results. One potentially significant imbalance is the rapid increase

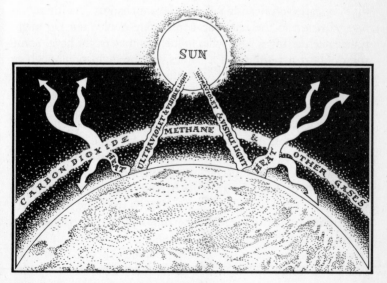

Heat trapped beneath a thickening layer of carbon dioxide and other gases could result in a global warming trend.

in carbon dioxide noticed by scientists in the past few decades. If the theories and computer models of scientists are correct, that build-up of CO_2 has the potential of dramatically altering the climate of the entire planet.

A greenhouse effect, or global warming, is created when particles of gas in the atmosphere occur in concentrations that allow heat energy from the sun to pass through but radiant energy from the earth to be trapped. The result is a net increase in ground-level temperatures. Even small increases have the potential to melt the polar ice caps, raise the oceans, and create widespread flooding of coastal regions. Changes in climate would have effects on life that can only be guessed at.

Until recently global warming has been primarily an abstract threat. Most scientists agree that atmospheric concentrations of carbon dioxide, methane, and other gases have increased by about 25 percent since the advent of the industrial age, and that those concentrations are continuing to increase. Virtually all agree that the gases trap solar heat in the famous greenhouse manner, and thus are capable of causing global climate changes. Yet, there has been division in the scientific community over when we might notice such changes, and what influences they might have. Some still doubt that warming, if it comes, will be dramatic enough or rapid enough to make much of a difference to life on the planet.

However, many biologists fear that global warming within the next fifty years will drastically affect the world's forests and wildlife. Scientists at NASA's Goddard Institute for Space Research predict that global temperatures will increase an average of 2.5 to 7.5 degrees Fahrenheit by the years 2040 to 2070, when carbon dioxide in the atmosphere has reached concentrations double those measured in 1980. Such an increase would make the atmosphere warmer than at any time in the last 100,000 years and would produce dramatic consequences. Alterations of forests, rising ocean levels, shifts in rainfall patterns, increasingly frequent and violent tropical storms, and other predicted results of global warming could well disrupt wildlife—and human life—all over the world.

Shooting Stars

When I heard the learn'd astronomer,
When the proofs, the figures, were ranged in columns
     before me,
When I was shown the charts and diagrams, to add,
     divide, and measure them,
When I sitting heard the astronomer where he lectured
     with much applause in the lecture-room,
How soon unaccountable I became tired and sick,
Till rising and gliding out I wander'd off by myself
In the mystical moist night-air, and from time to time,
Look'd up in perfect silence at the stars.

　　—Walt Whitman, "When I Heard the Learn'd Astronomer"

If Whitman stood long enough beneath that night sky, he probably saw meteors streaking overhead, a phenomenon that baffled learn'd astronomers until well into the nineteenth century.
The burning trails of meteors can be so vivid it is easy to understand why for thousands of years they were thought to be stars
falling from the heavens. Early sky watchers assumed that with
enough patience they could observe a particular star drop from its
place in the firmament, like a ripe apple falling from a tree. It
seemed a logical assumption in the Ptolemaic cosmological system,
where the earth was thought to be at the center of the universe,
surrounded by concentric spheres carrying the sun, the moon, the
planets, and the stars. It made sense too in ancient cosmological
schemes where the sky was an enormous cavern and the stars were
distant campfires. In Siberian legends the sky was a dome of sewn
hides through which the gods would occasionally peer, exposing

a flash of the radiance beyond. Some American Indians thought meteors were fragments of lunar material and called them "children of the moon."

Aristotle and other ancient Greek thinkers concluded that meteors were atmospheric, not astronomical, phenomena—thus the word meteor, devised from Greek sources meaning "high in the air," is the origin of our word "meteorology," or the study of the atmosphere. Pliny argued that meteors were celestial occurrences and offered a rather bizarre explanation for their appearance: "When the stars are believed to fall, what happens is that owing to their being overfed with a draught of liquid they give back the surplus with a fiery flash, just as with us also we see this occur with a stream of oil when lamps are lit."

Before meteors appear as burning streaks in the sky, they drift in space as bits of matter known as meteoroids. Some are fragments of the moon that were hurled into space millions of years ago by the impact of meteorites; others are pieces of planets. Some scientists have theorized that the fewer than 10 percent of recovered meteorites composed of iron and nickel are from the core of a planet that fragmented early in the history of the solar system, while those composed of rock originated in the crust of that planet. Others are fragments of asteroids left over after the formation of the planets, and are now orbiting the sun in a wide belt between Mars and Jupiter. The asteroids frequently collide, breaking off chunks and particles, some of which are thrown out of orbit by the gravitational pull of the planets and end up drifting close to earth.

Although meteors can originate from any of those sources, most of the 100 to 200 million visible meteorites that enter the earth's atmosphere each day are broken off, expelled, and scattered behind comets as they pass in orbit through the solar system. Comets, frequently described as giant dirty snowballs, are composed of frozen water, carbon dioxide, ammonia, and methane, and studded with bits of stone and ore. As a comet follows its long, elliptical orbit of the sun it gradually disintegrates, especially near the sun, where heat sublimates the comet's ice into vapor and unleashes rock, iron, and other debris that becomes strewn along the orbital path. When the comet's orbit intersects the earth's orbit those suspended bits of debris are attracted by earth's gravity and drawn into our atmosphere. Since the orbital path of a comet is relatively consistent, it is possible to anticipate with fair certainty which days of the year our own orbit will pass through it and produce meteor "showers."

When a meteoroid is captured by the earth's gravity it hurtles at great speed toward the ground. At an altitude of about sixty-five miles, having reached speeds of as much as 150,000 miles per hour —about 100 times faster than a rifle bullet—friction from gas molecules in the atmosphere heats the meteor until its outer atoms vaporize, leaving a trailing glow. The intensity and duration of the meteor's trail varies with its size and the direction of its approach. Generally the larger the particle, and the shallower its angle of descent (shallow angles cause it to travel slower, through a longer length of atmosphere), the brighter and longer the "shooting star" will appear. Most meteoroids are tiny, the size of bird seed or grains of sand, but occasionally large pieces enter the atmosphere and burn in brilliant, fiery displays.

A tremendous Leonid meteor shower over the eastern United States in November 1833 awakened both fear and scientific interest in meteors. During the early morning of November 13 the sky was filled with an estimated 200,000 meteors in six hours, all of them appearing to originate from a point in the constellation Leo. Many people thought it was evidence of the Apocalypse prophesied in Revelations ("and lo, there was a great earthquake, and the Sun became black as sack cloth, the full moon became like blood, and the stars of the sky fell to the Earth"). Astronomers were more interested to note that the Leonid meteor shower was apparently a cyclical event, occurring at about the same time every year as the earth's orbit intersected the meteor streams.

There remained much doubt about the origin of meteors until a carefully observed meteor shower in 1872 was determined to result from debris left over from the disintegrated comet Biela. Biela had been seen to split into two pieces in 1846 and to reappear as a double comet in 1852. After that it was never seen again, but when the earth crossed its orbit on November 27, 1872, there was a terrific meteor shower. The event showed for the first time a clear connection between meteors and comets.

It is a curious insight to human nature that meteors have been considered both a portent of evil and a sign of good luck. Wishing on a falling star is an old superstition with one important stipulation: The wish, if it is to come true, must be made while the meteor is visible. Unfortunately, most meteors fade far faster than a thought can be completed. For centuries in the British Isles it was customary to say a child had been born each time a meteor appeared, an idea that perhaps had its source in the biblical story of the Star of Bethlehem. Others have thought when a star fell it meant someone—often someone of importance, a "star"—had died. In *King Richard the Second*, Shakespeare has a Welsh captain warn the Earl of Salisbury, "And meteors fright the fixed star of heaven . . . These signs forerun the death or fall of kings." In legends of Central Asia, meteors were "fire serpents" coursing through the sky, sometimes bringing mischief, sometimes bringing loads of treasure. The Votiaks of western Siberia believed meteors to be "fire worms" that would descend at night to feed on the blood of sleeping mortals. To the people of the Andaman Islands, they were torches carried by evil spirits of the forest hunting for men.

Darkness is the key to successful meteor viewing. Watch on nights during the dark phase of the moon, or before or after the moon rises, from locations well away from city lights. Generally, the most meteors will be visible between midnight and dawn, when the dark side of the earth is facing forward in the direction of its orbit. Just as the front windshield of a car collects more raindrops than the rear windshield, the leading face of the earth is most likely to collect meteors. The trail, or "train," left behind by a meteor usually lasts less than a second. Those entering the atmosphere at a steep angle, directly at the observer, are extremely short and may appear as brief flashes or stationary bright spots. Those entering at a shallow angle can linger for several seconds, or, in rare instances, as long as a minute or a minute and a half.

Most appear white or yellow, but the longest and brightest can be red, orange, or even green.

On an average dark, moonless night, when visibility is good, expect to see five to ten meteors every hour. Numbers higher than that qualify as a meteor shower and occur on predictable dates throughout the year, when the earth crosses the trails of comets. The intensity of a meteor shower varies from year to year, with as few as ten per hour and as many as a few hundred per hour, but on rare occasions, a meteor shower can become an unbelievably spectacular meteor storm. On the night of November 17, 1966, for instance, during the Leonid meteor shower, some observers in New Mexico saw hundreds of meteors per second streaming through the sky. During one splendid forty-minute period, an estimated 100,000 meteors lit the sky in one of the greatest displays of meteors in recorded history.

Meteor showers are usually named for the constellation they appear to radiate from. That apparent radiant point is an illusion of perspective, like crepuscular rays at sunset, but it provides a handy reference system to identify individual showers.

Most years the brightest show of meteors is the Perseids, which usually peak during the early morning hours of August 12, when you can count on seeing an average of seventy-five meteors per hour. The rate is irregular, however, with lulls during which none are visible followed by periods of furious activity.

Meteor streams are not equally distributed, but consist of bunches separated by widely dispersed meteoroids. By analyzing historical accounts of Leonid showers from 902 to 1833, mathematician Hubert Anson Newton discovered that major falls of thousands of meteors per night occurred every 33.25 years, and he predicted—correctly as it turned out—one would occur in November 1866. It was a Leonid shower in 1933 that brightened the sky like "a child's sparkler," according to one witness. And thirty-three years later it occurred again in the 150,000-meteor-per-hour display over the western United States. By Newton's calculations we can expect a repeat performance in November 1999.

Fireballs and Bolides

WHEN LARGE METEOROIDS are trapped by the earth's gravity and drawn into our atmosphere, the resulting meteors are in a class of their own. They can become *fireballs*, which are usually defined as meteors brighter than the planet Venus, or *bolides*, which are exploding meteors. Pliny described a fireball over Rome in A.D. 66 that began as "a spark" that was "seen to fall from a star and increase in size as it approached the earth, and after becoming as large as the moon it diffused a sort of cloudy daylight, and then returning to the sky changed into a torch." A November 1977 fireball seen in parts of Ontario was described as being as bright as the sun. Witnesses in January 1983 in West Virginia reported hearing sonic booms and thunderlike rumblings three to five minutes after a fireball brighter than the full moon streaked across the sky.

Occasionally a large fireball passes through the atmosphere at such a shallow angle that it travels horizontally above the earth, then exits the atmosphere and "escapes" back into space. A well-known example observed by thousands of people, and recorded with a home movie camera and later shown on network television, was the fireball of August 10, 1972, which occurred in daylight over a 900-mile path from Utah to Alberta, Canada. Tracked closely by a U.S. Air Force satellite, it is known to have approached as close as thirty-six miles from the ground over Montana, where many witnesses heard sonic booms. Along the entire length of the fireball's route it streamed sparks and flashes and left a train like a jet's contrail that was visible for thirty minutes after it had passed. The meteor itself—probably a fragment of an asteroid—was estimated to be as long as a football field and to weigh a million tons.

Stones from Heaven: Meteorites

A FEW DOZEN times each year a meteoroid beats the odds against burning up in the atmosphere and falls to earth as a meteorite. Most meteorites began their descent as large meteoroids, especially fragments of asteroids (comet debris is usually too small and fragile to survive). Most enter the atmosphere at a shallow angle and at relatively low velocity.

Meteorites were long thought to have supernatural or divine origins and were considered worthy of reverence. It has been theorized that the Kaaba stone held sacred by the Muslims, as well as the stone worshipped in the temple of Diana in Ephesus during the third century B.C. were both large meteorites. In early ages, meteorites were valued for practical as well as religious reasons. Three types of meteorites have been recovered. Most are composed of stone, far fewer of stone mixed with iron, fewer yet of nearly pure iron (or about 90 percent iron and 10 percent nickel). It was long ago discovered that iron meteorites could be shaped into weapons and tools that were harder, stronger, and with keener edges than stone or bronze. By 1,500 B.C., when the Hittites of Asia Minor learned to smelt nearly pure iron from various ores by heating them with charcoal, they referred to iron as "fire from heaven." Earlier, the Egyptians had known iron as "stone of heaven," and the early Sumerian word for the ore could be translated as "sky" and "fire." Meteoritic iron was highly treasured as a gift of the gods thrown down from the heavens.

By the eighteenth century and the Age of Reason, many scientists in the western world considered meteorites a subject of folklore and mythology, and therefore not worthy of study. The notion that rocks might actually fall from the sky was absurd, as unlikely as talking sheep and flying cows. When the French Academy of Sciences commissioned chemist Antoine-Laurent Lavoisier in September 1768 to investigate reports of a "great stone" that had fallen from the sky near the town of Luce, he reported back that the

witnesses of that unbelievable phenomenon were either lying or mistaken. A large fall of meteorites on July 24, 1790, near Agen, in southwestern France, was witnessed and documented by more than 300 people, including the mayor of the town, yet was still declared an impossibility by the editor of a scientific journal who investigated the incident.

It was not until 1803, when the physicist Jean-Baptiste Biot investigated an enormous shower of 2,000 to 3,000 meteorites weighing up to about twenty pounds each near L'Aigle, about 100 miles west of Paris, that the Academy of Sciences was convinced.

Not everyone found the evidence incontrovertible. On December 14, 1807, a spectacular fireball streaked across the sky over New England, and crashed near Weston, Connecticut. Three hundred pounds of meteorites were collected by a Yale chemistry professor and the college librarian. When they submitted the stones as evidence that the French Academy has indeed been correct in giving meteorites a celestial origin, President Thomas Jefferson was said to remain skeptical. His often-quoted (and probably apocryphal) remark was, "It is easier to believe that two Yankee professors would lie than that stones would fall from heaven."

Indeed, those stones can fall. One that fell in Arizona 50,000 years ago left a crater 4,000 feet across and 750 feet deep. At least a dozen other large-impact craters known to have been caused by meteorites are found in Australia, Russia and the Baltic States, Saudi Arabia, Poland, Argentina, Texas, and Kansas.

The largest meteorite ever recovered is the Hoba iron, which was found in southwest Africa in 1920. It measures 9 x 9 x 3 feet, weighs an estimated 66 tons, and probably fell to earth in prehistoric times. Curiously, though partially buried, it is not surrounded by a crater.

Another notable meteorite is Ahnighito, a 34-ton iron specimen that fell in Cape York, Greenland, about 10,000 years ago. For centuries Eskimos used fragments of it to make knives and tips for their harpoons, until it was appropriated by Robert Peary who sold it in 1897 for $40,000 to the American Museum of Natural History in New York, where it is still on display.

The Kirin meteorite of China was viewed by thousands of eye-witnesses as it fell on March 8, 1976, near the outskirts of the city of Kirin. The main fragment buried itself eighteen feet in the ground and weighs about 3,900 pounds.

The Williamette, a 15.5 ton iron meteorite, was found in 1902 in Oregon by a farmer who spent a year hauling it 4,000 feet to his property from land owned by Oregon Iron and Steel. When he tried to sell the meteorite the Oregon Iron and Steel property reclaimed it, then sold it themselves to the American Museum of Natural History. It is now on display beside the Ahnighito.

Because most meteorites are nearly indiscernible from ordinary stones, it is difficult to find small specimens unless witnesses see them fall. That has happened only on rare occasions. On August 31, 1991, a thirteen-year-old boy named Brodie Spaulding was standing in his yard near Indianapolis when he heard a "low whistling noise" moments before a fist-sized, one-pound meteorite struck the ground just five feet from him. He reported the dense, heat-blackened rock was still warm when he picked it up.

Even closer encounters have occasionally been reported. A meteorite the size of a chariot was said to have struck and killed ten men in China in 616 B.C., and in separate incidents in the nineteenth century, a dog and a horse were reportedly killed by having the unimaginably bad luck of standing in the precise spot where a meteorite came to earth. On two occasions meteorites plunged through the roofs of houses in Wethersfield, Connecticut—once in 1971, again in 1982—without injuring anyone. There is only one documented case in recent history of a person being struck by a meteorite. On the afternoon of November 30, 1954, Ann Hodges, a thirty-two-year-old housewife in Sylacauga, Alabama, was bruised severely on the hip by an 8.5-pound meteorite (measuring seven by five inches) that crashed through her roof and ricocheted around her living room while she was taking an afternoon nap. The meteorite hit her on the rebound as she reclined on her couch.

The surest way to determine that an object is a meteorite is to find it in areas where no surface rocks exist—areas such as extensive ice fields. In 1969 a group of Japanese scientists surveying the immense glaciers of Antarctica discovered nine closely spaced rocks on the ice. They picked them up out of curiosity and carried them back to their base camp. Only later, after the rocks were examined closely, were they recognized as meteorites that had fallen over a period of thousands of years and been preserved in the glacial ice. Fresh meteorites continue to be uncovered as the ice is eroded by wind and weather, giving the impression that the region receives uncommonly heavy falls of them. In truth, meteorites fall with no more frequency in Antarctica than anywhere, but on the surface of the vast, featureless ice, the stones become easily visible. In what one scientist called "a big Easter egg hunt for grown-ups," scientists now gather as many as 1,000 meteorites and fragments of meteorites each year from the Antarctic ice fields.

Types of meteorites.

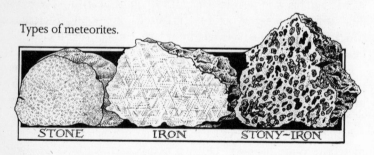

STONE   IRON   STONY-IRON

Major Meteor Showers

Date/Max	Name	Average/hour
Jan. 4	Quadrantids	50
Apr. 21	Lyrids	15
May 4	Eta Aquarids	20
June 20	Ophiuchids	15
July 25	Capricornids	15
July 28	Delta Aquarids	20
July 30	Pisces Australids	15
Aug. 5	Iota Aquarids	15
Aug. 11–12	Perseids	75
Oct. 20–22	Orionids	35
Nov. 1–4	Taurids	15
Nov. 16–17	Leonids	15
Dec. 5	Phoenicids	25
Dec. 13–14	Geminids	60
Dec. 23	Ursids	20

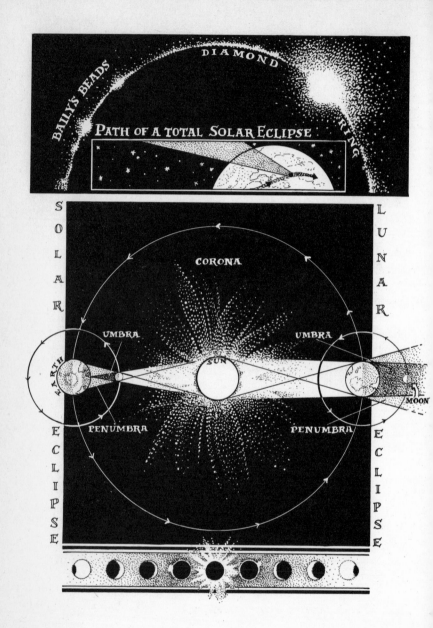

Blotting the Sun and Moon: Eclipses

In a world so utterly dependent on solar energy for life, what could be more terrifying than the death of the sun? To see it slowly devoured by darkness can stir the most elemental fears of extinction. Some anthropologists have suggested that humans first learned to pray during the trauma of eclipses, when desperate appeals to a higher power would have been rewarded with the return of the sun or the moon. But though their prayers and entreaties were answered, every civilization looked upon eclipses as bad news.

Like comets and meteors, eclipses seem random and chaotic, challenging all notions of the universe as an orderly place. The Romans were so convinced that eclipses were supernatural auguries of misfortune that it was considered illegal and blasphemous to say publicly that they were natural phenomena. The Roman naturalist Pliny had little patience for the "miserable mind of man" that sees "in eclipses of the stars crimes or death of some sort . . . or in the dying of the moon inferring that she was poisoned and consequently coming to her aid with a noisy clattering of cymbals."

To the ancient Chinese, Armenians, Maya, American Indians, and many other people, eclipses meant dragons, giant wolves, vampires, and other monsters were attacking and devouring the sun and moon. A Czechoslovakian legend tells of twelve ice giants, enemies of the sun, who occasionally conquered it, causing eclipses. If such monsters succeeded, the world would be thrown into perpetual darkness and the powers of evil would come down to devour humans. To fight off or frighten away these threats to the world, people built bonfires, shouted, threw stones in the air, banged pots, and fell to the ground wailing. The solar eclipse of

July 29, 1878, caused great distress among some American Indians. A report in the *Philadelphia Inquirer* said, "It was the grandest sight I ever beheld but it frightened the Indians badly. Some of them threw themselves upon their knees and invoked the Divine blessing; others flung themselves flat on the ground, face downwards; others cried and yelled in frantic excitement and terror. Finally one old fellow stepped from the door of his lodge, pistol in hand, and fixing his eyes on the darkened sun, mumbled a few unintelligible words and raising his arm took direct aim at the luminary, fired off his pistol, and after throwing his arms about his head in a series of extraordinary gesticulations, retreated to his own quarters. As it happened, that very instant was the conclusion of totality."

The oldest printed record of an eclipse is in the ancient Chinese text, *Chou King*, or the Book of History, which mentions an eclipse that occurred on October 22, 2136 B.C. According to the text, that eclipse caught the Chinese off guard, forcing them to assemble frantically for the usual rites—shouting, banging mirrors and gongs, beating drums, shooting arrows in the sky—to frighten away the dragon that was devouring the sun. The court astronomers, Hsi and Ho, had failed to prevent the eclipse because they were drinking rice wine instead of watching the sky for signs. Accused of negligence ("Stupidly, they went astray from their duties . . ."), they were beheaded.

The Chinese and Babylonians were among the earliest civilizations to notice that eclipses occurred with predictable regularity. They discovered that at intervals of eighteen years, nine to eleven days (varying because of leap years), and eight hours, the moon and sun repeated their orbital alignment relative to the earth. During that period, known as a *saros*, there were about forty-one total and partial solar eclipses—about four per year—and each eclipse occurred about 120 degrees west of the previous one. Any one spot on earth could thus be expected to see a total eclipse once every 400 years. The Maya left records in a document known as the Dresden Codex showing they were able to chart past eclipses and predict future ones with surprising accuracy. Herodotus told the story (sometimes challenged as apocryphal) of how the Greek mathematician Thales of Miletus announced a solar eclipse for the year 585 B.C. His prediction proved correct, and ended a war between the Lydians and Medes when the combatants interpreted it as a sign that they should cease fighting.

A knowledge of eclipses allowed Christopher Columbus to pry provisions from reluctant Jamaicans during his fourth voyage to the New World. When his request for food and water was at first denied, he consulted his navigation and astronomy tables, noticed that a lunar eclipse was due, and threatened to destroy the moon if supplies were not brought to his ship. The Jamaicans were dubious at first, but when the eclipse began as promised they quickly provided him with everything he needed. Mark Twain used the same idea in his novel, *A Connecticut Yankee in King Arthur's Court*, when his protagonist, a New Englander who, following a bump on the head, is transported to Camelot, makes a big impression on King Arthur and saves himself from being burned at the stake by successfully predicting a solar eclipse.

If the solar system was a more orderly place, eclipses would be regular and relatively unexciting events. Each month when the moon was new, it would cross in front of the sun and we would see a solar eclipse. Two weeks later, when the moon was full, it would pass into the shadow of the earth and we would see a lunar eclipse.

Eclipses are rare, however, because the plane of the moon's orbit around the earth is inclined about five degrees away from the plane of the earth's orbit around the sun. During the full and new moons, when the moon enters syzygy—falling into line with the sun and earth—it is seldom in perfect alignment. Most months the new moon passes slightly above or below the sun, and the full moon passes slightly above or below the shadow of the earth.

Occasionally the earth and moon reach positions in their orbits that put them in perfect alignment with the sun. When it happens on the day of the new moon or the night of the full moon, we see an eclipse.

The Sun Eclipsed

ONCE EVERY EIGHTEEN MONTHS or so, someplace on earth, day is turned inside out. Darkness sweeps across the world in a nearly palpable wave, leaving the land lit in a dim, unearthly glow. Temperatures drop, streetlights switch on, the wind dies. Flowers close, roosters crow, bats come out to feed. It is easy to understand why people who have no reason to believe eclipses are only temporary would be filled with fear and awe during the event. Even the most rational-minded observer can feel the stirring of those same emotions.

During a solar eclipse the intense shadow of the moon's umbra is drawn like a stylus across the planet, laying down a path only 80 to 100 miles wide. It travels at dizzying speed—the eclipse of March 7, 1970, drew a diagonal line across North America from Mexico to Newfoundland, a distance of 8,500 miles, in only three hours—and is visible as a total eclipse at any spot for only a few minutes. Outside that narrow track only a partial eclipse is visible.

It is a grand coincidence that the moon, about $\frac{1}{400}$ as large as the sun, is located about 400 times closer to the earth, making it the precise size to cover the sun during an eclipse. If the moon was smaller or farther from the earth, the eclipsed sun would shine around it in a brilliant ring; if it was larger or closer, eclipses would be darker, of longer duration, and more common. Still, there is some discrepancy. The moon's orbit is elliptical, meaning that during some eclipses it will be at its farthest distance from earth, at others its closest. When a solar eclipse occurs while the moon is at the far end of its orbit, it is called an annular eclipse—from the Latin *annulus*, for "a ring"—and the moon is not quite able to cover the sun's disk. When the moon is closer to the earth, it covers the sun entirely.

A solar eclipse begins slowly, with an hour or more passing between first contact and totality. Not until the moon has covered a substantial portion of the sun does the light become affected. When only a narrow crescent of the sun remains, rippling lines of light and dark called shadow bands—cast by irregularities high in the earth's atmosphere—sometimes cross the ground. Near totality, patches of sunlight may appear through the mountains on the horizon of the moon in a series of glittering spots called Baily's beads, named for the amateur astronomer who first discussed them in 1836. In the last moment before the moon covers the sun, the moon sometimes appears to be circled by a ring mounted with a brilliant jewel. This "diamond ring" phenomenon is caused when Baily's beads create a nearly continuous ring of light. The "diamond" is the tip of the sun just before it disappears behind the lunar horizon. If you are standing on a hill with a view to the west and south during that last shining moment, it is possible to see the shadow of the moon rushing across the landscape toward you. Then the sun goes out.

APHELION OF MOON ORBIT

ANNULAR ECLIPSE

During totality the landscape is thrown into a bizarre, nearly dark netherworld that is neither day nor night. Stars are visible in the sky, yet the sun's corona is bright enough to cast a dull, eerie light. Totality can last from two or three minutes to a maximum of seven minutes and forty seconds, depending on how close to the earth the moon is in its elliptical orbit.

In those few minutes of total eclipse the moon seems less an obstruction of the sun than a dark hole punched in the sky where the sun should be. Around that circular, black void shines the glimmering corona, a brilliant halo of burning gases in the sun's outer atmosphere. Pink shafts of the corona known as prominences thrust out from the fringes, sometimes extending more than five times the diameter of the sun—a distance of nearly five million miles into space.

A vivid description of a total solar eclipse was broadcast by CBS radio correspondent William Perry as he watched a spectacular eclipse in Peru in June 1937: "Darkness is really coming on us. It is an unhealthy and unnatural sort of dark. The brilliant light of the South Seas, the blues and greens of the sun, are now all as though washed down by dirty color. Darkness is now really upon us. We see the diamond ring—that famous circle around the moon. We see the sun streamers breaking through the mountains of the moon. We see for the first time the magnificence of the corona, which extends in great streamers away out."

Viewing the sun with unprotected eyes is always dangerous, whether during an eclipse or not. The safest way to watch an eclipse is to project it onto a screen. The easiest way to do that is to poke a pinhole in a piece of cardboard, hold it up between you and the sky (but do not look at the sun, even through the pinhole), and adjust its position until a projection of the sun can be cast on another piece of cardboard or paper serving as a screen. The same effect occurs naturally beneath a leafy tree: Hundreds of tiny images of the sun will be cast among the shadows on the ground and tree trunk. Or, make a projection box of cardboard, mount it on the viewing end of a telescope, and move the telescope until the image of the sun enters the lens and is projected onto the cardboard. Sunglasses are not dark enough to protect the eyes while looking directly at the sun, but a very dark welder's glass is sufficient (use glass number 14 or darker). Once the sun is completely covered during total eclipse it is safe to look at it with the naked eye, but use some caution. Even a small portion of sunlight flaring beyond the moon's horizon can damage the retina.

On average, one total solar eclipse occurs somewhere on earth every eighteen months, but there can be as many as two to five combinations of annular, partial, and total eclipses each year. In the final years of the twentieth century, total eclipses are due November 3, 1994 (visible in Peru, Brazil, and the South Atlantic); October 24, 1995 (visible in Iran, India, the East Indies, and the Pacific); March 9, 1997 (visible in much of Russia and the Arctic); February 26, 1998 (visible from parts of the Pacific and Atlantic oceans); and August 11, 1999 (visible in the British Isles, France, Turkey, and India). An annular eclipse will be visible over portions of the United States on May 10, 1994.

The Moon Eclipsed

THE LUNAR ECLIPSE of August 16, 1989, was unusual, not because lunar eclipses are great rarities—there are one or two most years—but because it took place on a cloudless, warm night. For my family and me it was unusual too because we were, for once, at the right place at the right time, sitting on blankets on the top of the highest of Lake Michigan's sand dunes, with a marvelously clear view of the sky in every direction. We watched the shadow of the earth first nibble at the edge of the moon, then move slowly across its face. Half-eclipsed, the shadow lowered across the moon like a heavy eyelid, and we could see the noticeably curved shadow of the earth, an observation that convinced Aristotle that the earth was round.

As the eclipse proceeded, the moon changed into shapes unlike any my sons, at ten and two years old, had ever seen. Aaron saw animal shapes: a duck's head, a snapping turtle, a snake with its mouth wide open around an egg, a rock singer with a Mohawk haircut, a Pac-Man character with its mouth frozen open. Nick, the two-year-old, watched in somber silence until the last sliver of sunlight laid across one edge of the shadowed moon, then announced in a sympathetic voice, "Moon's broke."

At full eclipse the night glowed with a peculiar coppery-red light. The moon was entirely blocked from the sun by the earth, yet it glowed with the bloody brightness of a full moon tinted red by clouds or urban smog. The coloration is due to atmospheric refraction—the bending of the red colors of the spectrum around the earth so that they, alone of all colors, reach the moon—and can cause the moon to appear brownish, deep-red, rust-red, brick-red, or orange. On rare occasions the eclipsed moon will appear entirely black, a result of large amounts of volcanic aerosols in the atmosphere. Black eclipses were reported after the massive eruption of Krakatoa on Java in 1883. In more recent times, they occurred after the eruptions of Agung in the South Pacific in 1963 and El Chichón in southern Mexico in 1982.

That night in 1989, the moon at full eclipse was copper-colored and bright enough to be clearly outlined against the night sky; even its dark seas were visible. The center, less luminous than the edges, gave it dimension, like an orange. Around us, the dunes, the few trees, and the scattered clumps of people sitting on blankets were illuminated in the strange light.

Totality lasted a little more than an hour. The entire duration of the eclipse was much longer, but the early and late stages, when the moon passed through the penumbra, did not alter the appearance of the moon. If the moon had passed through the precise center of the umbra, total eclipse could have lasted a maximum duration of one hour and forty-five minutes.

Lunar eclipses are visible over large areas of the earth, but not everywhere because in some locations the time at which it passes through the umbra can occur before the moon rises or after it sets. Most people will be able to observe total lunar eclipses on December 9, 1992; June 4, 1993; November 29, 1993; April 4, 1996; September 27, 1996; September 16, 1997; January 21, 2000; and July 16, 2000.

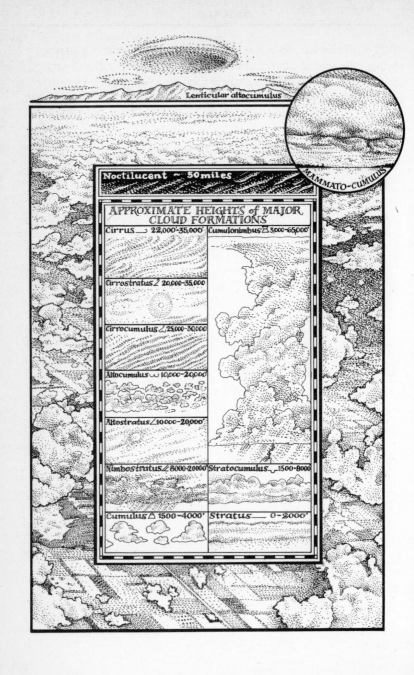

Lenticular altocumulus

MAMMATO-CUMULUS

Noctilucent ~ 50 miles

APPROXIMATE HEIGHTS of MAJOR CLOUD FORMATIONS

Cirrus — 22,000'-35,000'   Cumulonimbus △ 3,000'-65,000'

Cirrostratus ∠ 20,000-35,000'

Cirrocumulus ∠ 25,000-30,000'

Altocumulus ⌒ 10,000-20,000'

Altostratus ∠ 10,000-20,000'

Nimbostratus ∠ 8000-20,000'   Stratocumulus — 1500-8000'

Cumulus △ 1500-4000'   Stratus — 0-2000'

IN THE CLOUDS

Thoreau compared an unclouded sky to a meadow without flowers and a sea without sails. Clouds obscure the sun, they bring rain and snow, they cover the world in a gray cloak, but without them the pure blue sky of daylight becomes as uninteresting as a canvas without paint or a page without words.

As early as the eighth century B.C. Ionian philosophers proposed that clouds were a form of thickened, wet air. That idea passed through many variations over the centuries, reaching perhaps its most peculiar interpretation in the writings of an Italian priest, Urbano d'Aviso, who wrote in 1666 that water vapor was composed of "little bubbles of water filled with fire, which ascend through the air as long as it is of greater specific gravity than they are; and when they arrive at a place where the air is equally light, they stop." Most scientists of the eighteenth century, including Edmond Halley, believed that water vapor consisted of hollow spheres. Even as late as the middle of the nineteenth century, the British writer and critic John Ruskin believed the apparent buoyancy of clouds could be explained by "hollow spherical globules" within the clouds, "in which the enclosed vacuity just balanced the enclosing water." Ruskin had not kept up with the scientific findings of his time. By the turn of the nineteenth century the English chemist John Dalton was developing the theory—essentially correct, as it turned out—that water vapor is a gas and that it behaves in the air in much the same way as oxygen, nitrogen, or carbon monoxide.

In 1803, Luke Howard, a thirty-year-old amateur meteorologist from England, published a classification system that for the first time gave names to types of clouds. Howard classified clouds according to three primary shapes: heaps of separate clouds with flat bottoms and bulbous tops, which he named *cumulus* (Latin for "heap"); layered clouds that formed a blanket-like covering, which he called *stratus* ("layer"); and wispy, curling clouds which he named *cirrus* ("fiber" or "hair-like"). He gave the name *nimbus* ("rainy") to those with heavy moisture content that resulted in precipitation, and combined terms to describe clouds that included several of the major features. Thus, the towering, anvil-topped cumulus that produce major thunderstorms became "cumulonimbus," and thin, wispy clouds that suspended in definite layers became "cirrostratus."

The system found immediate widespread acceptance. Today, meteorologists recognize ten major types of clouds found within three basic altitudes. Low-altitude formations, found at heights up to about 10,000 feet, include cumulus, stratocumulus, and stratus. Cumulus, scattered like fat sheep across a blue meadow, are often called "fair-weather cumulus," and rarely result in precipitation. Stratocumulus are dark, dense, somewhat ominous, and usually cover the entire sky; they appear checkered or wavelike, sometimes with breaks of open sky between them, and usually produce only localized, short-lived showers of rain or snow. Stratus are less dense, tend to be white and wispy, and look like high-drifting fog; they might produce drizzle but seldom rain.

Among the middle-altitude formations, at heights of 10,000 to 20,000 feet, are altostratus, altocumulus, and nimbostratus clouds. Altostratus often appear as a uniform opaque covering through which the sun or moon shines weakly; they often trail fallstreaks or virga—a fine descending drift of ice crystals and water droplets that evaporate before reaching the ground. Altocumulus are patchy small clouds separated by breaks of open sky, usually found at heights around 10,000 feet; they may produce localized rain or snow showers. Nimbostratus are the dark, seemingly impenetrable clouds associated with all-day rains and serious snowfall; they usually cover the entire sky, with no discernable base, and result from vast masses of warm, moist air.

High-altitude formations are found from 20,000 to 50,000 feet high, and can be identified by their thin, wispy, wind-swept appearance. At those heights are found cirrus, cirrostratus, and cirrocumulus clouds, none of which ordinarily produce much rain or snow, although they are often precursors of other, precipitation-bearing clouds. Cirrus clouds are high, thin, and white, and are often blown into elaborate, curling formations by upper level winds; composed of ice crystals they sometimes precipitate, but the ice and snow that falls from them often evaporates before reaching ground. Cirrostratus are thin, white, and veil-like, composed of ice crystals, and are usually found at heights above 20,000 feet; they are the clouds most often responsible for lunar and solar haloes, and are often an indication of denser cloud cover to come. Cirrocumulus are among the highest of clouds, usually found above 25,000 feet, and are composed of small convection cells that give them the appearance of fish scales—the "mackerel sky" of many old weather proverbs; they are an indication of unstable air and often precede rain.

Towering formations of cumulonimbus can drive through all three levels, sometimes reaching to the very top of the troposphere. Cumulus that rise in the morning, especially on hot, sunny, summer days, by afternoon often grow into high formations with cauliflowerlike shapes (sometimes labeled swelling cumulus and cumulus congestus), and frequently release brief rain showers. These growing towers of cumulus rise from the low into the middle altitudes, reaching heights up to 30,000 feet and are evidence of unstable and energetic atmospheric conditions. With enough heat and moisture they become cumulonimbus—thunderheads—with enormous tops that can reach ten or twelve miles high. Cumulonimbus clouds are the spawning grounds of severe thunderstorms, high winds, and tornadoes. The tops of these massive formations become flattened and anvil-shaped when they reach the tropopause—the ceiling of the troposphere—where stable conditions and warmer temperatures create an inversion that prevents the cloud from growing any higher. The top of the thunderhead often becomes wispy or feathery in appearance as ice crystals within the top of the cloud are drawn out by upper-level winds.

A number of unusual cloud formations combine some of the features of the main types, or are formed when wind arranges clouds into striking shapes. My grandfather used to tell me that when the clouds look like upside-down pans of biscuits, run for cover. He was referring to mammatocumulus or mamma, which form as ranks of pouches hanging from the underside of storm clouds. They are caused by pockets of cold air descending from the cloud mass, and sometimes appear shortly before tornadoes develop in the area. They also occur when particularly severe thunderstorms are beginning to disintegrate.

Various unusual clouds form at the tops of mountains. Among them are banners and caps, created when moisture-laden air is driven up the slope of a mountain and condenses at low temperatures. Others, called lenticular altocumulus, are saucer-shaped—so clearly saucerlike, in fact, that they are sometimes reported as UFOs—and form at the crest of waves in the jet stream above mountain peaks.

The highest of all clouds are noctilucent, or "luminous night clouds." These high, thin, streaky clouds occur only rarely, and are noticed usually because they remain brightly lit long after sunset or before sunrise, giving an indication of their immense heights. Most occur in the northernmost latitudes of Europe, Canada, and Alaska, and usually only in late summer. Oddly enough, they are always located about fifty miles above the ground, an altitude where the air temperature is a consistent −100 to −130 degrees Fahrenheit, and where water vapor should be virtually nonexistent. The origin of noctilucent clouds has been argued in the past, but most scientists now agree that they are usually caused by small amounts of water vapor that have penetrated above the stratosphere and were converted instantly into clouds of ice particles. A few are created artificially by the exhaust of rockets launched through the upper atmosphere. Fewer yet are the residue of the short-lived but visible trails of large meteors as they burn up at the fifty-mile level.

The Anatomy of a Cloud

CLOUDS ARE ESSENTIALLY condensed water vapor, but it takes more than an accumulation of water to form them. Temperature is also important: Warm air can absorb much more water than cold air—five times more at 80 degrees then at 32 degrees—which is why you can see your breath on a winter day and not on a summer day. In cold air moisture condenses readily into visible clouds of droplets. Warm air saturated with water vapor rises, cooling as it goes, until it reaches a level where it can no longer hold the vapor. This height, called the condensation level, is where clouds are formed. Because the temperature level tends to remain fairly constant, the clouds tend to stay at about the same level.

Even air heavily saturated with moisture will not condense unless there are airborne particles for water molecules to cling to. Dust, pollen, plant spores, smoke, salt from ocean spray, and volcanic ash can all serve as condensation nuclei. Such particles are microscopic in size, usually smaller than the dust motes you can see drifting through a shaft of sunlight in a still room. In laboratories, a flask of moist, dust-free air will condense instantly into fog when a cloud of tobacco smoke is blown into it.

In the atmosphere, a similar effect is seen when a high-flying aircraft passes overhead. The jet trail, or contrail, is an artificially induced cirrus cloud, formed by water vapor as it is ejected from the jet's engines condensing onto the constant stream of tiny particles left by the aircraft's exhaust. Jets typically fly at heights where the air temperature is about −70 degrees Fahrenheit. Such cold air is quickly saturated by only a small amount of moisture. If the air contains even a moderate amount of water vapor a contrail will linger in the air for hours; if the air is dry the contrail dissipates quickly. Lingering contrails tend to swell and drift until they look like strips of cirrus clouds. At sunset they become as brilliant and colorful as any other cloud, and can create unusual effects. They are rare at sunrise, since most jet traffic occurs during the day, and the previous day's trails will have dissipated during the night. Contrails can offer clues to upcoming weather. If a jet leaves no trail or if the trail fades quickly, it is an indication of a relatively dry upper atmosphere, which suggests the weather will remain stable and fair. Long-lasting contrails stretching across the entire sky are evidence of high moisture content in the upper atmosphere and the approach of a warm front and subsequent precipitation.

Clouds are never static. They are in constant movement from the wind and from the continuous loss and gain of the individual water droplets that form as saturated air comes in contact with the existing cloud. Shifting, drifting, constantly diminishing and replenishing themselves, clouds stay suspended in the air because of the lifting action of air currents directed upward by obstructions like hills and mountains, and especially by convection—warm air currents rising from the surface of the earth. They contribute to the very winds that move them, reflecting sunlight away from the earth, so that the ground directly below is cooler than unshielded ground. Air rushes to achieve equilibrium between these pockets of uneven temperature. When the sun sets and its direct heat no longer contributes to convection, rising currents of air quickly lose their force. Cumulus clouds in particular are subject to nightly descent as air currents cool, yet instead of descending all the way to the ground and enveloping the surface in fog, the clouds as they descend warm and compress their water droplets, causing them to evaporate. The clouds literally disappear before they reach the ground. This phenomenon, most noticeable when illuminated by the full or nearly full moon, has long been noted, and inspired an old French saying, "The full moon eats clouds."

Watched with the perspective of time-lapse photography, clouds are revealed as surprisingly predictable and orderly. Fair-weather cumulus, those chubby, white, roughly spherical clouds that so often dot the sky on summer days, are not the random and casually drifting things they appear to be at a glance. Each cloud is born on a current of heated air, or a thermal, rising from cultivated fields, parking lots, and other surface features that attract and hold heat from the sun. The heated air rises like a spouting fountain, carrying invisible water vapor until it reaches temperatures cool enough for the vapor to condense, forming a cloud. Once formed, the cloud presses upward, moisture at the top evaporating while new moisture at the bottom condenses, and develops within itself bulbous lobes known as "florets," each blossoming and fading and disappearing during a life span of ten to fifteen minutes. Meanwhile, cool air from the top of the cloud spills down around its borders, creating a space of clear air around each cloud. That vertical movement creates wind, which merges with horizontal winds and keeps the cloud moving. As it leaves the warm, saturated column of rising air it disintegrates, and another begins forming in its place. The life of a cumulus cloud, alas, is often brief.

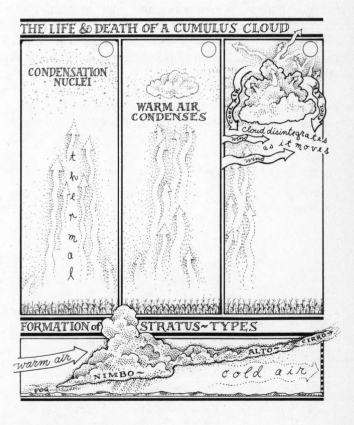

THE LIFE & DEATH OF A CUMULUS CLOUD

CONDENSATION NUCLEI

thermal

WARM AIR CONDENSES

cloud disintegrates as it moves

COOL AIR

wind

wind

FORMATION of STRATUS~TYPES

warm air

CIRRO~

ALTO~

NIMBO~

cold air

FOG

White clouds are composed of small water droplets that scatter the colors of sunlight. Uniformly scattered light distributes all the colors of the spectrum equally, which our eyes always see as the color white. Storm clouds and dense formations of stratus appear gray or black because water droplets within them are much larger than the droplets contained in fog or fair-weather clouds, and absorb more light, thus appearing a dark color. "Old" clouds can also be gray or purplish, because most of the small droplets have evaporated out of them and the few large drops remaining are not as efficient at scattering sunlight.

Surprisingly, in spite of the abundance of clouds in the atmosphere at any one time around the world, they account for only a minute amount of the earth's moisture. If all the water vapor contained in the atmosphere—whether visible as clouds or invisible as uncondensed vapor—were to condense suddenly and fall to the ground, it would produce only about an inch of rain worldwide.

If you're hoping for rain, don't place too much hope in a single cloud: A typical fair-weather cumulus cloud a few hundred yards in diameter contains only about twenty-five gallons of water.

WHERE THERE'S THUNDER, THERE'S LIGHTNING

Few weather events make us so aware of the power of nature as thunder and lightning. Our efforts to understand and control the weather, to build enduring monuments, to shape the world according to our whims all seem trivial in the face of a romping thunderstorm. The world is untamed and untameable. Wind, rain, and lightning beat us down to size.

Thunderstorms originate in cumulonimbus clouds, or thunderheads, when the following conditions exist: a large area of moist, unstable air; little or no wind (which stirs the atmosphere and prevents overheating of air); an initial impulse—like the heat of the sun—to begin the process of convection. Once those conditions are met, heat from the surface of the ground warms and expands the surface layer of air, causing it to rise quickly. The heated air cools as it rises, and moisture condenses out of it, which in turn releases more heat, allowing convective air to ascend as high as four or five miles. Think of it as an enormous, slow-motion explosion of moist air. It results in strong vertical winds and massive cloud formations with cauliflowerlike summits ballooning into the shape of mountains.

Inside thunderheads, by a process that is not yet fully understood, electrically charged ions separate, negative going to the bottom, positive to the top. Positive and negative ions are always present in the atmosphere, but they are usually paired up in nearly equal proportions. Most scientists suspect that the violent vertical winds within a thunderstorm cause particles of water and ice and molecules of air to electrify, gaining or losing electrons and separating them into powerful charged fields. Air is a poor conductor of electricity, and serves to keep the charges apart for a time. But

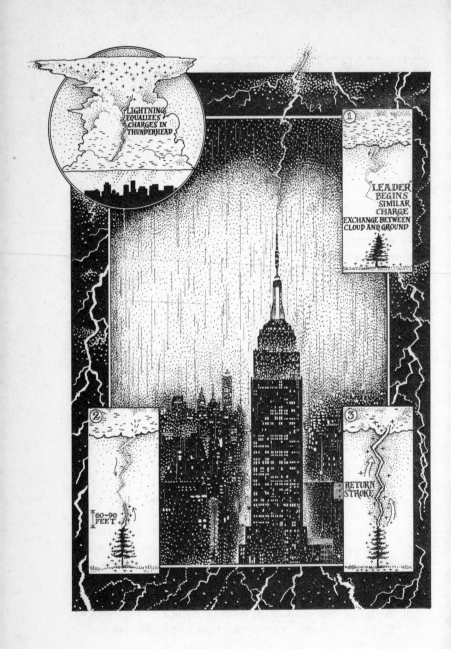

LIGHTNING EQUALIZES CHARGES IN THUNDERHEAD

① "LEADER" BEGINS SIMILAR CHARGE EXCHANGE BETWEEN CLOUD AND GROUND

② 60~90 FEET

③ RETURN STROKE

when the difference in voltage between the top and bottom of the cloud becomes too great for the air to resist, a bolt of lightning equalizes the charges by leaping the gap within the cloud, like a spark jumping from an electric outlet to a plug.

At the same time electric fields are forming within a cloud, a similar disparity is building between the bottom of the cloud and the earth. Like the atmosphere, the earth is ordinarily charged with both positive and negative ions. When the negative charge at the base of a thunderstorm passes overhead, however, it attracts a predominance of positive charges to the ground and in the air just above it, in a kind of electrical shadow that follows the storm cloud as it moves. The charge can be powerful enough to cause the hairs on your neck and arms to rise into the air, or for mild electricity to flicker and glow from the uppermost points of high objects. Again, air serves as an insulator, but only until the opposing charges in the cloud and on the earth build to an irresistible level.

At that moment something remarkable happens. Until a decade or so ago it was assumed that charges were equalized between the cloud and the ground by a single, blazing bolt of lightning. Slow-motion photography showed, however, that a lightning strike is far more complicated than that. It begins with the discharge of a faintly visible, relatively low-powered "leader" that emerges from the cloud and shoots toward the earth in a series of 150-foot, zigzag steps, each step lasting only about fifty-millionths of a second. The real fury of lightning explodes when the leader is sixty to ninety feet from the ground. It is met then by an upward-seeking discharge of positive ions blasting skyward in a return stroke of electricity two or three inches in diameter and surrounded by a four- or five-inch sleeve of superheated air. The stroke packs 10,000 to 200,000 amperes and creates temperatures of 50,000 or more degrees Fahrenheit, instantly cooking the surrounding air, causing it to expand violently in a roar of thunder. At the moment the return stroke enters the cloud another leader forms and follows the same path down to the earth, is met by another rising charge, and returns again to the cloud. The process repeats itself three or four times (and as many as twenty-six times), but the bolts travel up and down so fast—up to 93,000 miles per second—our eyes see only a single, blinding flash of lightning.

The Roman poet and philosopher Lucretius, who lived from about 100 to 55 B.C., theorized in *De rerum natura*, *The Nature of the Universe*, that thunder was caused by the collision of clouds: "First, then, the reason why the blue expanses of heaven are shaken by thunder is the clashing of clouds soaring high in the ether, when conflicting winds cause them to collide." The less charming explanation offered by modern meteorologists is that lightning's sudden, intense heat causes the air along its path to expand abruptly, creating sound waves that radiate outward. When lightning strikes very close, you hear a nearly instantaneous clap of thunder followed by rumbling peals. The rumbling is the sound of thunder being produced as the return stroke climbs to the cloud, exploding the air on the way up. Short bolts of lightning create correspondingly short claps of thunder. If the bolt reaches across a great distance, or if it occurs between high hills that create an echo, or if a number of bolts flash in quick succession, the thunder "rolls."

THUNDER

RADIATING SOUND WAVES CAUSED BY EXPANSION OF SUPERHEATED AIR ALONG PATH OF LIGHTNING BOLT

In the United States thunderstorms are most prevalent in July and occur more frequently in Florida than any other state. They are a way of life in Kampala, Uganda, where they appear on average 242 days each year. An even more thunderous place was Bogor, Indonesia, where from 1916 to 1920 it stormed at least once a day for 322 days per year. An estimated 2,000 thunderstorms are in progress at any given moment somewhere on earth—44,000 per day, 16 million per year—and lightning strikes the earth approximately 100 times per second. A thunderstorm with a diameter of three miles might contain 500,000 tons of water and a potential energy equal to ten atomic bombs like those dropped over Hiroshima and Nagasaki.

Thunder and lightning have inspired rich mythologies everywhere, and are often portrayed as weapons of the gods. Thunderbolts of justice have been hurled by such varied deities as the thunder-bird of African tribes, the Egyptian god Seth, the Greek god Zeus, the Roman god Jupiter, the Norse god Thor, and the Old Testament god Jehovah. In Africa and ancient Rome people killed by lightning were thought to have deserved divine wrath and were buried hurriedly, on the spot, without ceremony. Prehistoric axe-heads and meteorites were "thunderstones," and considered theological proof that the gods could smite mortals who displeased them. Hundreds of tales tell of people being struck by lightning at the moment they uttered blasphemies, or otherwise challenged the gods. The Roman poet Virgil wrote of an audacious prince named Salmoneus, who claimed he was a god and drove his chariot over a bridge of bronze to create thunder and tossed torches along the way to produce lightning—and was promptly struck and killed by a bolt from heaven as punishment for his hubris. Sometimes divine justice has to strike hurriedly to make its point. In the nineteenth century a murderer in Kentucky was reported to have been struck and killed by lightning as he stepped onto the scaffold to be hanged.

Few cultures have been as serious about lightning as the Romans. For centuries major affairs of state were determined by the College of Augurs, a three-member group of sky-watchers who observed and interpreted the significance of lightning, birds, and meteors. The Augurs wielded enormous power, since no one had the authority to question their version of celestial happenings. If they said lightning flashed across the sky from right to left, clear evidence that Jupiter did not approve of the current state of government affairs, then all meetings of the public assembly were canceled and recent decisions were reconsidered.

Superstitions about lightning have thrived for centuries. Pliny wrote, "When in fear of lightning men think caves of greater depth are the safest, or else a tent made of the skin of the creatures called sea-calves, because that alone among marine animals lightning does not strike, just as it does not strike the eagle among birds . . ." He noted furthermore that there are some kinds of thunder, "which it is not thought right to speak of, or even to listen to." Rural Britons also once considered it bad luck to talk about lightning, or to count the seconds between lightning and thunder, or even to point in the direction from which it was expected. Children guilty of such dire offenses were sometimes made to kneel blindfolded on the floor to get an idea of how it would feel to be blinded by the lightning they had tempted. Scandinavians believed that a Christmas Yule log would protect a house from lightning all year long. In England a piece of hawthorn wood cut on Holy Thursday was thought to make a house and its inhabitants immune. Similarly, a house could be protected by supplying it with sprigs of holly, mistletoe, St. John's wort, houseleek, and hazel that had been gathered on Palm Sunday. Oak trees were considered by some a protection against lightning, by others an attractor of it; it was generally believed that wood from an oak that had been struck by lightning was a useful charm. Throughout Europe it was considered good insurance during a thunderstorm to hide the scissors, cover every mirror, stay away from wet dogs and horses, and lie on a feather bed.

Medieval Europeans routinely inscribed church bells with the Latin words *Fulgura frango*—"I break the lightning"—and as late as the eighteenth century would ring the bells at the first sign of a thunderstorm, believing the holy bells would disperse evil spirits or emit sound waves with the power to repel lightning from its downward path. Regardless of the rationale, the effort often proved futile. A German text published in 1784 under the reasonable title, *A Proof That the Ringing of Bells During Thunderstorms May Be More Dangerous than Useful*, reported that during a thirty-three-year period a total of 386 church towers were struck by lightning, and 103 bell-ringers were killed on the job.

Like the Romans centuries before them, English folk of the Middle Ages agreed it was good luck to hear thunder on the left (though others preferred it on the right). In sixteenth-century England it was believed that thunder on Sunday would cause the death of learned men and judges; on Monday, the death of women; on Tuesday it was evidence of a bountiful grain crop; Wednesday, the death of harlots; Thursday, proof of a healthy flock of sheep; Friday, the murder of a great man; and Saturday, general pestilence and plague. A modern Maryland folklorist noted that a toothpick carved from a tree struck by lightning would cure toothaches—an idea advocated by Pliny in A.D. 77.

Thunderstorms occur primarily in summer, but on rare occasions they have been reported in winter, and even during snow storms. Perhaps because it is so unusual, many Europeans long considered winter thunder an ill omen. A seventeenth-century text pronounced it "ominous, portending factions, tumults, and bloody wars, and a thing seldome seen." A popular adage of the same era said "Winter's thunder is summer's wonder," or, more ominously, "Winter thunder bodes summer hunger." In Wales it was believed that winter thunder would presage the death of the most important person within a twenty-mile radius.

The old notion that lightning never strikes the same spot twice is mostly a case of wishful thinking. The Empire State Building is struck about forty times each year, and the bronze statue of William Penn on Philadelphia's City Hall has been struck too many times to count. When I was a child I had friends whose grandfather survived being hit by lightning three times. He was struck on two occasions while walking between his house and barn, and the third time while riding his tractor in a corn field. That third strike impressed me most, because I had always been told that a rubber-tired vehicle was safe from lightning (it isn't), and because the impact threw the old man off the tractor to the ground and left him with a baseball-sized bruise on the top of his thigh, where the lightning bolt had struck him, and a matching bruise on the opposite side, as if it had passed straight through.

My friends' grandfather is not the only person known to have attracted more than his share of lightning. The champion of all such survivors was probably Roy C. Sullivan, a former National Park ranger from Waynesboro, Virginia, who was struck by lightning seven times in his life, suffering chest and stomach burns, the loss of a toenail, the singeing of both eyebrows, and flames (twice) in his hair. Sullivan, who died in 1983 of natural causes, was struck while fishing, while driving a truck, while sitting in a fire tower, while walking through a campsite, and while standing in his front yard. Once, during a storm, he took cover inside a ranger station, but even that did not help: Lightning struck the building, followed electric lines inside, and jumped from an outlet to zap him. Asked by a reporter in 1977 why he was so attractive to lightning, he said, "Lordy, I wish I knew. It's awful."

Bead Lightning, Ball Lightning, and St. Elmo's Fire

A JAGGED STREAK forking from the bottom of a cloud is the most familiar form of lightning, but certainly not the only one. On rare occasions the lightning bolt will appear to be composed of a string of luminous balls, like a chain of beads. This so-called "bead lightning" appears to be an interrupted stroke, as if the ionized channel forged by the leader stroke is incomplete.

In February 1799, Captain Haydon of Her Majesty's Ship *Cambrian*, recorded in his journal: "Observed a tremendous squall coming down upon us; turned the hands up to clew up the close-rigged topsails. While they were so employed a ball of fire struck the topmast-head, killed two men and wounded many others. The number taken below amounted to about twenty."

Captain Haydon's crew were probably struck by a rare and controversial form of lightning known as "ball lightning." Various eyewitness accounts have described the balls varying from one-half inch to six feet in diameter, shaped as spheres or ovals, and colored white, red, yellow, or blue. They have been described as barely moving or streaking along at great velocity, and they are often said to bounce when they strike the ground. Sometimes, after a few seconds or a few minutes, they burst with a loud explosion. It is not universally accepted that ball lightning even exists, since it has never been studied in controlled circumstances, but a number of reliable eyewitnesses have described the phenomenon. One was a British meteorologist named J. Durward, who had two memorable encounters with ball lightning. The first was in Scotland in the summer of 1934, when he and his son watched a ball of fire about a foot in diameter drift toward them from a grove of pine trees. It struck an iron gate post while Durward's son was touching the latch, and gave him a shock severe enough to make his arm numb for hours afterwards. The second encounter took place four years later during a thunderstorm over France in a flying boat. Durward and the pilot watched a ball of fire enter the open cockpit window, pass so close to the pilot that it singed his eyebrows and burned holes in his seat belt, then drift to the rear of the cabin and explode. Similar eyewitness accounts are not uncommon, but little is known of the nature or cause of the phenomenon.

When the negative charge at the bottom of a thunderstorm attracts positive ions on the earth beneath it, they tend to congregate at the peaks of the highest available objects. That static charge sometimes illuminates in a pale, blue, hissing glow the tops of trees and buildings, the ice-picks of mountaineers, the wings of airplanes, and the rigging of ships. English sailors in the nineteenth century called such lights *corposants*, a mispronunciation of *corpo santo*, "the body of the saint," but it has become better known as St. Elmo's Fire. Named in honor of the patron saint of sailors in the Mediterranean, it was long considered a good omen by mariners, perhaps because it often occurs after the worst of a storm has passed.

Although St. Elmo's Fire may indicate a lightning strike is imminent, the radiance itself is usually considered harmless. Nonetheless, it has been blamed for igniting the fire that destroyed the airship "Hindenburg" on May 6, 1937, at Lakehurst, New Jersey.

Heat Lightning and Sheet Lightning

THE DISTANT, silent flash of heat lightning on the horizon is so common on summer nights that it seems a unique feature of the season, quite different from the noisy, jagged bolts of thunderstorms. But heat lightning—sometimes known as summer lightning—is nothing more than ordinary lightning reflected in clouds far in the distance. It seems silent because at a distance of five to ten miles the waves of sound generated by thunder become too widely dispersed to be heard by human ears. We see the fury, softened by the miles, but hear no sound. Thoreau watched distant, soundless flashes of lightning and wrote that the clouds seemed to "lift their wings like fireflies."

Sheet lightning occurs deep within clouds, and appears from a distance to illuminate large areas uniformly, like a flashbulb set off inside a tent. But it is ordinary lightning, arcing from the positive-charged upper portion of the cloud to the negative-charged lower cloud. Because it never leaves the cloud we see no jagged bolt.

Lightning on the Wing

AIRCRAFT ARE STRUCK in flight an average of once per year per airplane, but their metal shells usually conduct the electricity harmlessly from one end to the other and back out into the air. Occasionally, flight instruments are affected or the crew may be blinded temporarily by the flash, but lightning-caused crashes are rare.

Birds unlucky enough to be struck by lightning seem to fare worse than airplanes. The journal of the American Ornithologist's Union, The Auk, reported that in 1941 four double-crested cormorants were seen struck by lightning and killed as they flew over a field in South Carolina. The large and broad-winged pelican seems peculiarly prone to lightning. On August 16, 1929, a gas station attendant was watching a flock of twenty-seven white pelicans pass 500 feet overhead near Great Salt Lake, Utah, when lightning struck and killed the entire flock, scattering their bodies over ten acres of land. Ten years later, in 1939, a farmer near Nelson, Nebraska, watched thirty-four white pelicans rain to the ground after a bolt of lightning zigzagged through their flock. More recently, in April 1990, sixteen migrating pelicans were struck and killed as they passed over Broadwater, Nebraska.

When Lightning Strikes

LIGHTNING IS BEAUTIFUL and fascinating—and deadly. Between 100 and 300 people are killed by lightning each year in the United States, and as many as 1,500 are injured. Fatalities from lightning usually exceed those from tornadoes, hurricanes, and floods combined.

The safest place during a thunderstorm is inside a building, away from windows, and out of contact with telephones, plumbing, and electrical wires. Outside, you are safest inside a car with the windows rolled up. If caught outdoors without shelter stay away from solitary trees, wire fences, wet beaches, and water, and never raise golf clubs or umbrellas in the air. Taking cover in a thick stand of small trees is much safer than being the tallest object in a field or on a hill. If you are in the open, with no shelter available, crouch with your hands around your knees to keep the least possible contact with the ground.

Where there's thunder, there's lightning, but because sound

travels much slower than light (1,088 feet per second—or about one mile in five seconds—as opposed to light's 186,000 miles per second), the lag we hear between the two grows with distance. To calculate your proximity to a bolt of lightning, count the seconds between the flash and the next peal of thunder, then divide by five. Thus, a lag of fifteen seconds indicates the lightning flashed about three miles away.

Nature's Tantrums: Tropical Cyclones

Of all the great storms that rip across the surface of our planet, none can match the magnitude and power of the great ocean-spawned tantrums meteorologists know as tropical cyclones. They are found in every tropical ocean, affect every continent except Antarctica, and have been the cause of some of the worst natural disasters in human history.

Although an essentially planet-wide phenomenon, tropical cyclones are isolated enough to have earned different names in various parts of the world. In the eastern North Pacific, Caribbean, and Atlantic they're known as hurricanes, a word derived from the Taino *huracan*, for "evil spirit," and known in the Caribbean as the God of All Evil, who would send terrifying winds to punish people when he was angered. The same storms in the Pacific are called typhoons, a mispronunciation of the Chinese *ta-feng*, for "violent winds." In Australia they are sometimes called willy-willies, while elsewhere in the Southern Hemisphere and in the Indian Ocean they are known as cyclones. The word *cyclone* made its first appearance in print in Henry Piddington's *Handbook for Sailors* in 1844, the term coined by him to suggest that the great circular storms of the Indian Ocean had a shape similar to a coiled snake.

Aristotle considered hurricanes a combination of winds piling upon other winds until they generated into great storms of enormous power. Lucretius wrote that when a powerful storm wrapped itself in clouds and burst upon the land, "it belches forth a whirlwind and storm of enormous violence; but as it seldom takes place at all and as mountains cannot but obstruct it on land, it is seen more frequently on the sea with its wide prospect and unobstructed horizon."

TROPICAL CYCLONE

60,000 FEET

EYE

FAVORED ORIGIN AREAS

For centuries, the great storms that raked the oceans and terri-fied mariners were considered isolated, local events. No one had the means to see the bigger picture, to recognize that the tempest that sank a fleet of English ships off Barbados was the same storm that days later destroyed villages on the Bahamas. The first accurate description of the workings of a tropical storm did not appear in print until the end of the seventeenth century, when the English explorer and buccaneer William Dampier published a treatise ti-tled, Discourse of Trade-Winds. Dampier compared a typhoon in the China Sea to an enormous, extremely violent whirlwind, and said that the "Tuffoons" that sometimes battered the coast of China were no different than the hurricanes that periodically swept across the Caribbean. Not long after Dampier's book was pub-lished, an amateur scientist named William Redfield noticed that trees and buildings knocked over when a hurricane struck the New England coast in 1821 also suggested a gigantic whirlwind, with a pattern that revealed the winds within the storm had moved in a counterclockwise direction.

Tropical cyclones are almost always born in the summer or fall, in areas over the oceans between latitudes 5 and 20, north and south of the equator. They are the product of a complex arrange-ment of circumstances, but have their primary source in the mid-summer heating of tropical oceans. When temperatures are high enough to heat surface water to 80 degrees Fahrenheit or higher, thunderstorms over the warm water rise high into the air and cause the atmospheric pressure at the surface of the ocean to drop. Meanwhile, trade winds in the area begin spiraling counterclock-wise in the Northern Hemisphere and clockwise in the Southern Hemisphere as a result of the earth's rotation. Warm, moisture-laden winds pulled toward the center of the spiral are drawn upward by the low pressure at the center, begin to increase in velocity, and concentrate the spinning of the storm system. When winds reach thirty-nine miles per hour the storm becomes classi-fied as a tropical storm and is given a name. If the winds surpass seventy-four miles per hour, as they do in approximately one out of ten tropical storms, the storm is upgraded to a tropical cyclone.

If conditions are right, it can grow to more than 300 miles across, progress at a speed of 10 to 50 miles per hour, and generate internal winds of 100 to 200 miles per hour.

The most peculiar feature of a hurricane is the eye, a five to twenty-five mile circular nucleus of calm, where dry air at high altitudes is forced downward through the center of the storm. Survivors of severe hurricanes sometimes talk about the strange experience of winds of unbelievable strength ceasing suddenly as the eye passes over, and of being able to look up at pure blue sky surrounded by a roiling mass of angry wallclouds spiraling tens of thousands of feet in the air. Ships have been known to ride out the worst of a hurricane by matching the speed of the hurricane and staying within the eye. In that comparative calm the riggings and

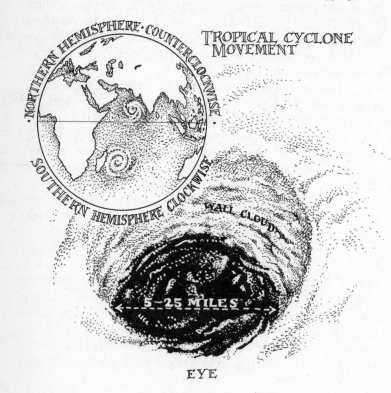

NORTHERN HEMISPHERE·COUNTERCLOCKWISE

SOUTHERN HEMISPHERE CLOCKWISE

TROPICAL CYCLONE MOVEMENT

WALL CLOUD

5-25 MILES

EYE

decks of vessels sometimes become crowded with exhausted birds seeking refuge.

The lifespan of a tropical cyclone is typically about ten days, during which it follows a generally western path, but one which is so capricious and erratic that it is difficult, if not impossible, to predict accurately. The storm is able to gain and maintain its enormous power for such long periods of time because it draws its life from the ocean. As it passes over water it picks up a continuous source of energy from the transfer of heat from the water to the air. Warm air rushing toward the center of the storm rises quickly. As it rises it expands and cools, forcing condensation. Since one of the side effects of condensation is the release of heat, the air is heated even as it rises, forcing yet more condensation and increasing the power of the updraft feeding the storm. If the storm travels north over colder water or if it passes over land, it loses the warm, moist air that feeds it, and will usually dissipate fairly quickly.

Most severe weather, even potentially disastrous storms like tornadoes, can be appreciated if seen from a safe distance. But the only distance sufficient to appreciate tropical cyclones is from hundreds of miles above, the altitude of a satellite's view, where a cyclone drifting harmlessly over an ocean looks like a whorl of smoke shaped in the delicate spiral pattern of a snail's shell. Up close there is nothing harmless about a hurricane, and anyone who is close enough to experience it firsthand will rightfully be more concerned with surviving it than watching it.

What is it like to be inside a hurricane? "Good God!" Alexander Hamilton wrote after witnessing one in the Caribbean. "What horror and destruction—it is impossible for me to describe—or you to form any idea of it. It seemed as if a total dissolution of nature was taking place. The roaring of the sea and the wind—fiery meteors flying about in the air—the prodigious glare of almost perpetual lightning—the crash of the falling houses—and the ear-piercing shrieks of the distressed were sufficient to strike astonishment into Angels." The Polish-born English novelist Joseph Conrad, who weathered many tropical storms in his years as a ship's captain, described the experience in *Typhoon*:

. . . a ragged mass of clouds hanging low, the lurch of the long outlines of the ship, the black figures of men caught on the bridge, heads forward, as if petrified in the act of butting. The darkness palpitated down upon all this, and then the real thing came at last.

It was something formidable and swift, like the sudden smashing of a vial of wrath. It seemed to explode all round the ship with an overpowering concussion and a rush of great waters, as if an immense dam had been blown up to windward. In an instant, the men lost touch of each other.

The most devastating feature of tropical cyclones are not their winds, but the storm-surge waves that often accompany them. The intense low pressure surrounding the hurricane, contrasted with the relatively high pressure at the eye, creates a suction on the surface that raises a hill of water one to ten feet high in the center of the storm. The bulge pushes water ahead as the storm moves across the sea like an enormous, blunt-nosed barge. The waves build as they progress, magnified by the winds, rising as high as twenty-five feet in the open ocean, even higher where the shores of a bay or channel compress them. When the storm surge coincides with high tide it raises the level of the ocean many feet, causing it to rush far inland. The combination of high winds, flooding from torrential rains, and storm surges have been responsible for more human deaths throughout history than earthquakes and volcanoes. Their grisly track record includes a forty-foot-high wave in the Bay of Bengal in 1737 that drowned 300,000 people; an 1881 storm that killed 300,000 along the coast of China; 500,000 deaths on the coastal islands of Bangladesh (then known as East Pakistan) in 1970; and another 125,000 killed in the same area in April 1991. The worst natural disaster in U.S. history was a hurricane that in 1900 killed more than 6,000 in Galveston, Texas.

Destructive as they are, hurricanes serve a few useful purposes. They moderate global temperatures, helping dissipate the intense heat of the tropics and transport it to latitudes where it is appreciated and carrying much-needed rainfall to regions that might otherwise get very little precipitation.

The tradition of giving feminine names to hurricanes was begun by military meteorologists during World War II. Starting in 1953 each year's storms were named alphabetically, with the first tropical storm and hurricane of the season beginning with the letter A, the second with letter B, the third with letter C, and so on. After 1979, the sexist assumption of the system was dropped, and since then female and male names have been used alternately.

SUNRISE, SUNSET

Our days begin and end with beauty. At midday the sun is a presence too powerful to ignore but too brilliant to look at, and at night we only know it second-hand, reflected off the moon and planets. Only at sunrise and sunset are we able to observe openly the star that makes life on earth possible.

We know the sun to be a middling-sized star, measuring about 865,000 miles in diameter (109 times larger than the earth), radiating heat from a center where hydrogen is being converted to helium by nuclear reactions that generate temperatures up to 40 million degrees Fahrenheit. In the scale of stellar things, it is middle-aged, some 5 billion years old, halfway through its life expectancy of about 10 billion years.

Civilizations of every degree of sophistication seem to have understood the central, life-giving role of the sun in our lives. The Egyptians personified the sun as Re, the Greeks as Helios, the Romans as Sol, the Aztecs as Tonatiuh; all considered it a god of supreme importance. Pliny, in *Natural History*, gave a tribute to the sun that is a curious mix of primitive and classical attitudes:

> In the midst of these [the planets] moves the sun, whose magnitude and power are the greatest, and who is the ruler not only of the seasons and of the lands, but even of the stars themselves and of the heaven. Taking into account all that he effects, we must believe him to be the soul, or more precisely the mind, of the whole world, the supreme ruling principle and divinity of nature. He furnishes the world with light and removes darkness, he obscures and he illumines the rest of the stars, he regulates in accord with nature's precedent the changes of the seasons and the continuous rebirth of the year, he dissipates the gloom of heaven and even calms the storm-clouds of the mind of man, he

lends his light to the rest of the stars also; he is glorious and pre-eminent, all-seeing and even all-hearing.

To best appreciate this all-seeing and all-hearing preeminence it is necessary to rise early in the morning or wait until late in the day. This much at least of what your mother told you is true: Staring at the sun can make you go blind. Only when it is shining lengthwise through the atmosphere at sunrise and sunset are enough harmful rays deflected to make it safe to view the sun with unprotected eyes. Those same miles of atmosphere routinely create some of the most colorful scenes on earth.

Even after witnessing sunrises and sunsets all our lives a particularly vivid display can stun us to silence. Pliny thought the colors so vivid they could only be caused by fire: "It has often been seen, and is not at all surprising," he wrote, "that the sky itself catches fire when the clouds have been set on fire by exceptionally large flames."

Those flamelike colors are in fact caused by light passing horizontally through the atmosphere. The lower the sun, the greater the mass of air it must shine through, and the more of its light is scattered by particles. Straight overhead, enough light is scattered by the atmosphere to make the sun—which shines with white light—appear yellow. When it is low in the sky, shining through a greater mass of atmosphere, most of the shortwave colors are scattered away, leaving red the predominant color. Since the atmosphere is composed of many layers of air, representing different temperatures and differing amounts of water vapor, dust, ash, and gases, the light is affected in many ways. That explains not only why the sun changes colors near the horizon, but also why it often appears distorted. Sunlight near the horizon is subject to a great deal of refraction, or bending, as it passes through the atmosphere. Refraction occurs most noticeably at sea level, where the atmosphere is densest, and can cause the sun to appear as much as its full diameter higher than its true position. Another effect of refraction is the apparent flattening, or oblateness, of the sun (or the moon) as it nears the horizon. It becomes more elliptical than circular, sometimes stretching so far it separates into strips or steps.

The sun and sky appear red primarily because atmospheric gases scatter the blue light of the color spectrum more strongly than red light, allowing more red than blue to penetrate through to an observer. Generally, on hazy days with low visibility, you can count on a spectacular sunset; on clear days be resolved to a more modest show. The colors are enhanced by clouds; under the best conditions the clouds are thick enough to reflect the red colors but not thick enough to obscure the sun.

Since much of the spectacle of sunsets and sunrises is played out on clouds, their presence or absence usually decides how much color is visible. Yet even an unadorned sky can be stunning. Much of the color of an unclouded sunrise or sunset is visible before the sun rises or after it sets, when light is refracted high in the atmosphere to create zones of subtle and delicate color. When the sun is low it saturates the sky with color—deepening the blue overhead, lighting bands of yellow, yellow-green, and violet near the horizon, tinting the sun itself reddish-yellow. The closer it is to the horizon the darker and more intense those colors become. Reddish-yellow deepens to orange, then to red-orange, then to red. When the sun is below the horizon, reds blend with blues to spread shades of pink and light violet. Through twilight—from sunset until dark—the bands of color condense and shrink, and the light gradually fades from the sky. The old rule of thumb was that twilight ended when you could no longer read a newspaper by the light of the sky, but astronomers are more inclined to say it ends when the sun is 18 degrees below the horizon and the first stars become visible.

In the French Alps, Alpenglow is produced on clear, cloudless days when the subtle colors of the sun, though gone from the valleys, tint the snow-covered peaks of the mountains. Of course the Alps have no monopoly on the phenomenon: Alpine glows are common on mountains throughout the world.

A few minutes after sunset, turn toward the eastern sky, opposite the sun, and if the sky is unclouded but slightly hazy it is sometimes possible to see the earth's shadow appearing in a curving, bluish-gray band just above the horizon. In a fringe at the top of the shadow will frequently be a faint band of red or pink light known as the Belt of Venus, where the atmosphere remains in sunlight and is colored by the light's horizontal passage through the air. The greater the amount of haze in the sky, the more the atmosphere acts like a screen with an image projected on it, and the more visible the shadow will be.

It is also possible after sunset or before sunrise to see the zodiacal light—a curving or pyramidal glow in the sky caused by sunlight reflecting off tiny dust particles high in the atmosphere. In the morning this light is sometimes called false dawn, and in equatorial regions it will occasionally stretch across most of the sky. It is called zodiacal light because it appears in that part of the sky known as the zodiac—the ecliptic, or imaginary "belt" of the heavens where the sun, moon, and visible planets pass.

If you are a great fan of each day's beginning and ending moments, you might consider living near the ends of the earth. The duration of sunsets and sunrises varies radically according to latitude. At the equator, the sun rises and sets quickly because it travels at or near a 90-degree angle in relation to the horizon, and thus drops quickly out of sight. The closer you get to the poles, the less direct is the sun's route, and the longer it takes to pass over the horizon. Thus twilight lasts only a short time in the tropics, yet goes on for hours in the Arctic.

Dedicated sunset and sunrise watchers know that one of the best-known weather proverbs—"Red sky in morning, sailors take warning; red sky at night, sailors' delight"—often proves an accurate forecast. That seems paradoxical, because red sky at night might indicate cloud cover approaching with the prevailing winds, therefore bringing rain and wind, while red sky in morning could mean those same clouds have passed to the east and are being followed by better weather. What often happens, however, is that red sky at night occurs when the sky to the west is clearing, allowing the sun to break through the tail end of a weather front, and colors are reflecting on clouds that are already in the process of disintegrating. In the morning, an approaching rain front becomes a screen on which the rising sun casts its colors, and often means the day will be rainy.

Volcanic Sunsets

VOLCANIC ERUPTIONS send enormous quantities of fine ash high into the stratosphere, where it is dispersed around the world. In the first year or two after a major eruption, the sky hundreds or thousands of miles from the volcano will appear hazy or milky. The sun will be surrounded by a large white or lightly colored patch known as an aureole, or, less commonly, by a faint brownish pink band with a pale green interior, known as a Bishop's ring. First described by Reverend S. E. Bishop, who noticed the phenomenon in Hawaii in 1883 after the eruption of Krakatoa near Java, a Bishop's ring is sharply defined, with a hollow interior filled with silver or silver-blue light, and is markedly different from a solar halo, which is caused by refraction in ice crystals.

Some of the world's most unusual sunsets and sunrises have occurred in the weeks, months, and years following eruptions of gases and dust into the atmosphere. Observers throughout the world commented on the particularly vivid sunsets visible after the August 1883 eruption of Krakatoa. The eruption sent an estimated four and a half cubic miles of stone into the air at a velocity of one half-mile per second, killed 63,000 people, and thundered so uproariously it could be heard on the island of Madagascar, 3,000 miles away. For months afterwards, sunsets around the world were breathtaking. They were so brilliant on the evening of October 30 in New Haven, Connecticut, and Poughkeepsie, New York, that fire alarms were sounded and fire brigades mobilized because it was thought huge fires were burning to the west. The March 1963 eruption of Mount Agung on the island of Bali, in the South Pacific, and the 1980 eruption of Mount Saint Helens likewise resulted in extraordinarily brilliant and colorful displays.

Clouds of volcanic ash tend to remain suspended at an altitude of about 65,000 feet, and to be lit for about forty-five minutes after sunset and before sunrise. They are more widely dispersed than noctilucent clouds, and create a larger, more general display of colors.

The eruption of Mount Pinatubo in the Philippines in June 1991 was one of the largest in the twentieth century. Within six months ash from that eruption had completely circled the world, creating a slight thinning of the ozone layer and some temporary global cooling. Sunsets and sunrises following Pinatubo were the fieriest since Mexico's El Chichon in 1982 and Indonesia's Mount Agung in 1963. In late November we could see the effects in northern Michigan, where clear-sky sunsets were tinged in vivid and unusual hues of violet long after the sun was down.

Under normal conditions, when there have been no volcanic eruptions for several years, clear sky twilights are mostly blue and gray, with yellows and oranges appearing only near the horizon. A volcanic sunset, on the other hand, adds red to everything. Normal yellows and oranges grow brighter and expand to cover a larger area of sky. The regions above, which would normally be dark blue, become washed with red, mixing to become purple, purplish-pink, salmon, bright red, or brilliant scarlet. The primary glow band, that strip of orange and yellow light low in the sky when the sun is below the horizon, will be brighter, larger, and longer-lasting than usual. The sun itself can appear coppery, blue, or even green, shining out of a background of red, violet, gold, or silver. Ash clouds, sometimes called ultracirrus, are occasionally visible just after sunset as wispy striations high in the atmosphere, as if the sky had been brushed with a very weak gray wash of watercolor, leaving the brush strokes visible. They remain illuminated well after dark because of their high altitude.

Volcanic ash causes twilights to be prolonged much longer than usual. Light is scattered over a wide area by the veil of particles and colors can be visible as long as thirty minutes after the end of astronomical twilight. Ash high in the atmosphere can also cause the moon and stars to appear green when seen through the last of the purple twilight. The moon will sometimes be tinged blue as its light is refracted through the ash.

The effects produced by volcanic ash in the atmosphere tend to be periodic. A week or two of brilliant sunsets will be followed by a period of normal skies. Gradually the frequency and brilliance of the displays fade as the atmosphere cleanses itself of the particles.

The Mysterious Green Flash

FOR MORE THAN A century, sky observers have argued about the cause of a brilliant green flash of light that sometimes occurs for a fraction of a second just before the sun rises or just after it sets. The flash appears only rarely, usually when the atmosphere is very clear, and only when the sun is outlined against a distinct horizon, such as a mountain ridge or ocean. Jules Verne described the phenomenon in fiction, mentioning an ancient Scottish legend that attributed supernatural powers to the flash: "At its appearance all deceit and falsehood are done away, and he who has been fortunate enough to behold it is enabled to see closely into his heart and to read the thoughts of others."

There is still some debate over the cause of the green flash, but it is almost certainly a type of refraction phenomenon. In their book, *Sunsets, Twilights, and Evening Skies*, Aden and Marjorie Meinel of the Optical Sciences Center of the University of Arizona argue convincingly that atmospheric refraction causes the image of the sun to be separated into a ring of colors, with green located slightly above the predominant red image. For a second or less, when the sun is just below the horizon, a sliver of the green band is refracted above the horizon and becomes the last bit of visible light, erupting in a silent, instantaneous flash of emerald. The flash is just as likely to occur at sunrise as sunset, although it is more difficult then to be watching the precise spot where the sun will appear.

Rays of Light and Shadow

A COMMON FEATURE of sunrises and sunsets are the crepuscular rays that radiate outward from the sun like splayed fingers of light. They can be explained fairly simply as light made visible by dust and other particles in the atmosphere. Their apparent convergence is an illusion of perspective, like railroad tracks that appear to join together in the distance.

Sometimes it is not the rays of light but blue-gray shafts of shadow known as cloud rays that radiate away from the sun. Most common during the hazy, stagnant-air periods of summer, they are the shadows of clouds near the horizon that act to break the sun's light into rays and beams.

An effect similar to crepuscular rays occurs sometimes when the sun is high in the sky and shines downward through openings in clouds. The bright, well-defined beams when descending into oceans or lakes have been described as the sun "drawing water," as if the heat of the sunbeam evaporated water and carried it off into the atmosphere. The phenomenon is also known in some parts of the world as "Jacob's Ladder," a reference to the biblical story of Jacob's vision of angels ascending and descending a ladder reaching from heaven to earth. Scientists describe it as an example of the Tyndall effect, named for John Tyndall, a nineteenth-century British physicist who discovered that when light travels through dust or smoke the light is scattered by the particles and reflected to an observer's eyes, making the beam of light visible. The beams that break through cloud cover are made luminous by dust and water vapor in the atmosphere, and are enhanced by the darkness of the surrounding clouds.

Spots on the Sun

AT SUNRISE and sunset it is occasionally possible to see the solar surface marred by sunspots. To call them mere spots is incredibly misleading: They are often larger in diameter than the earth. First noticed by Galileo and his contemporaries—many of whom were skeptical of sunspots' existence, since the idea of a flawed sun suggested a flawed God—sunspots have since been recognized as regions of magnetic storm on the surface of the sun. They appear dark because they are cooler than the surrounding surface.

Although the results are anything but conclusive, there have been countless studies on the effects of sunspot activity on the earth. The numbers of sunspots vary according to an eleven-year cycle. During high activity, when vast amounts of charged protons are blasted in all directions by magnetic storms, our atmosphere is effected in various ways. The charged particles increase radio interference, for instance, and produce brilliant and more frequent auroras. Some studies have found a connection between sunspot cycles and population cycles of marine algae, coral colonies, fish, insects, and some mammals. Less easy to demonstrate are the attempts to blame sunspots for increases in human accident rates, epidemics, viral illnesses, and heart disease.

Unholy Illusions: Mirages and Other Optical Phenomena

In 1818, the British explorers John and James Ross were searching the frigid waters between Greenland and Baffin Bay for a passage between the Atlantic and Pacific oceans, when they found their way blocked by mountains looming above the sea ahead of them. Discouraged, they returned to England and reported that no Northwest Passage existed. Robert Peary met the same obstacle in those waters seventy-five years later and likewise decided the route was blocked. He called the range of mountains Crocker Land and went elsewhere in search of clear sailing. In 1913, an expedition headed by Donald MacMillan set out to explore Crocker Land, and was bewildered to find the mountains 200 miles west of the position reported by the Rosses and Peary. MacMillan and his companions walked on foot across miles of pack ice in an effort to reach shore, but they seemed to make no progress. The faster they walked the faster the mountains receded before them. It was not until the midnight sun dipped briefly below the horizon that the astonished explorers saw the jagged peaks of Crocker Land dissolve to nothing. The mountains that had blocked the Arctic from exploration were a mirage.

As air near the ground settles into layers of different temperatures, strange things happen. Light behaves differently in different mediums, bending, for instance, when it passes through hot air into denser cool air or from air into water—which is why the legs of someone standing in a lake or swimming pool appear bent at such an unnatural angle. When light shines on the boundary between two mediums like hot and cool air it might as well be

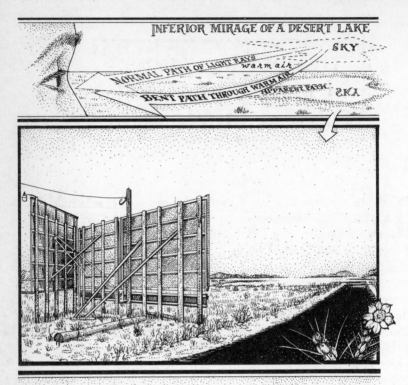

INFERIOR MIRAGE OF A DESERT LAKE

SKY

NORMAL PATH OF LIGHT RAYS

warm air

BENT PATH THROUGH WARM AIR — APPARENT PATH

SKY

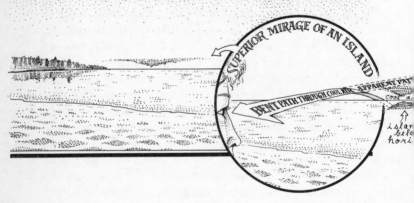

SUPERIOR MIRAGE OF AN ISLAND

BENT PATH THROUGH COOL AIR — APPARENT PATH

cool a

island
belo
hori

bouncing off mirrors. By the time an image crosses a broad expanse of those reflective air surfaces, nothing an observer sees can quite be trusted.

The most common mirages are probably the shimmering puddles that appear on highways during hot days. That patch of glistening water on the road ahead is an example of an inferior mirage—inferior because objects are caused to appear below their actual position—and is actually a reflection of the sky. The reflection is caused by a layer of hot air collected by the dark surface of the road from the sun. Light rays near the ground are reflected upwards and capture the image of the sky. The same thing happens in a desert, where palm trees seem to grow upside down on the horizon and vast lakes appear to shimmer in the distance. Puddle-mirages on hot pavement are harmless, but in the desert, where water is absolutely crucial to life and mirages are a cruel joke known to have lured travelers to their deaths, Arabs call them *Bahr el Shaitan*—Lakes of Satan.

Mirages of another sort occur when layers of cold air settle next to the ground. Standing on the frozen ice of a large inland lake I was once startled to see a dozen colorful ice shanties along the far shore hanging in the air, ten or fifteen feet above the ice. It was an example of a superior mirage, caused by light being bent downward by the dense layer of cold air on top of the ice. The downward-angled light makes it possible to see objects that are below the horizon, or, as in the shanties on the frozen lake, to see objects lifted higher than their actual positions. In the oceans near the north and south poles, frigid water and air heated by twenty-four-hour days combine to create very high, distinct temperature stratifications. When a layer of warm air is located far above the cold air near the water, reflected light will sometimes cast an image high above the sea. It is not uncommon to see an enormous iceberg hanging inverted, hundreds of feet in the air. The survivors of Scott's South Polar Expedition in Antarctica in 1912 saw a double image of their supply ship—one inverted, the other upright, the smoke from their smokestacks merging in the middle—suspended in the sky while the coast was still obscured by a range of large hills. Similar conditions have made it possible to see the inverted image of Iceland's mountains when they are still hundreds of miles away across open ocean, causing some writers to speculate that it was just those types of mirages that enabled Viking mariners to navigate to North America in their primitive ships.

On occasion temperature layering of the air will cause distant objects to appear grotesquely magnified, elongated, and much taller than they are. One form of this towering or looming effect is known as the Fata Morgana, named for King Arthur's sorceress half sister, Morgan le Fay, who was fond of conjuring castles in the air. Fata Morganas are extremely rare, created when complex and uneven arrangements of layered air creates superior and inferior mirages simultaneously. A layer of cold air sandwiched between two layers of hot causes images to reflect, build upon one another, multiply, tower, and grow until an immense, spectacular mirage hangs in the distance. In Italy's Strait of Messina, one of the few places in the world where the complicated mirage is seen with some regularity, the spectacle transforms a simple cloud on the horizon into a beautiful harbor town, complete with white-gowned pedestrians and colorful shops, with palaces rising in elaborate layers topped with ramparts and towers. Some observers claim they have seen details of harbor towns hundreds of miles away reflected in the Fata Morgana. Others claim it is the ghostly specter of cities long ago swept away by the sea.

Halos, Coronas, and Glories

LIGHT REFRACTING through water droplets and ice crystals in the atmosphere can produce a dazzling variety of optical phenomena. The most common of them are solar and lunar *halos*, formed when light is separated into its component colors by the prism effect of the six-sided ice crystals collected in a thin covering of cirrostratus clouds. Light rays are bent by the ice at an angle of 22 degrees; when they converge at the eye of an observer he sees a 22-degree ring—about the diameter of two fists held at arm's length—around the sun or moon. The halo is a pale, faintly colored ring with red on the inside and blue on the outside.

The same or similar conditions can produce various rings, arcs, spots, or patches of color, all known collectively as halo phenomena. Several partial and complete solar halos are sometimes visible at the same time, giving the sun a target and bull's-eye effect, or branching off with symmetrical "legs" and "wings" from the central halo.

If the cirrus clouds are intermittent they can produce partial halos called *parhelia* (from the Greek for "with the sun"), also known as sun dogs and mock suns. They probably became known

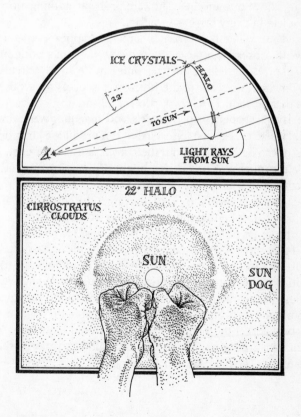

as sun dogs because they sometimes have a long horizontal ray of white light sticking out at one end like a dog's tail. The resulting bright patches are similar in appearance to the sun through a thick layer of clouds, and are most common when the sun is low in the sky, shining through loose cirrus. A halo will often have sun dogs on either side of it, lined up in triplicate with the sun. They can be hazy white or rainbow colored.

Sun pillars are caused when enough ice crystal surfaces reflect a beam of sunlight upward, like a stationary spotlight from the rising or setting sun.

A *double sun* is an uncommon halo effect that appears when the true sun's double hangs just above it (or, less often, below it). On extremely rare occasions there will be two or more double suns lined up above one another. They are probably nothing more than incomplete but very bright sun pillars.

A *subsun* is visible only from an aircraft. It appears in ice clouds below the observer, at an altitude below the sun's position, and looks like a long ellipse that becomes more nearly circular the higher the sun rises above the horizon. On occasion it can be so bright that it produces its own halo—a subsun dog.

When light shines through thin clouds containing water droplets of similar size rather than ice crystals it can produce a lunar or solar *corona* (not to be confused with the corona at the outer edge of the sun's atmosphere). A corona is a bright disk of pink or white light in a continuous disk, usually with a pink or red outer rim. The smaller the droplets in the cloud, the larger the corona will appear.

An odd halo phenomenon sometimes occurs in early morning when the sun is just up and a closely trimmed lawn is soaked with dew. To observe it, stand with your back to the sun, so that your shadow is cast across the damp grass. If conditions are right a luminous white halo will surround the head of your shadow, creating an effect eerily reminiscent of the halos said to surround the heads of saints. Thoreau noticed it near his cabin on Walden Pond: "As I walked on the railroad causeway, I used to wonder at the halo of light around my shadow, and would fain fancy myself one of the elect." The halo, known as a *dew halo* or *heiligenschein* (German for "holy shining"), is startling enough to have caused the immodest sixteenth-century sculptor Benvenuto Cellini to imagine he had been sainted, or at least given divine reward for his genius.

Unfortunately for Benvenuto, dew halos, while certainly not an everyday occurrence, can appear over the heads of saints and sinners alike. They form when sunlight, streaming past a viewer's head, strikes dewdrops. In much the same way that rainbows are formed, light penetrates each drop, is bent slightly, then reflects off the rear of the drop back to the viewer.

A similar phenomenon frightened the climbing shorts off early mountaineers who reached the fog-shrouded peak called Brocken, in Germany's Hartz Mountains. Frightened climbers returned from the mountain and told of a bizarre apparition they had seen climbing along with them near the peak. The stories were quickly added to ancient legends of the Brocken peak of the Hartz Mountains as the place where witches gathered on Walpurgis Night, a legend Goethe used in the witches' sabbath scene in *Faust*.

The apparition, dubbed the "Brocken specter," proved to be the climbers' own shadows enlarged and cast on clouds or fog banks above, below, or beside them. The dispersal of light in water droplets within the fog sometimes causes the shadows to be ringed with a colorful *glory*, or "Brocken bow," draping the specter in rainbowlike colors and adding to the eerie effect.

Glories are refraction phenomena most commonly seen when you stand facing mist or a bank of fog with the sun behind you, and your shadow is cast against the white screen of the fog. The glory appears around the head of your shadow as a series of colored rings. They are commonly seen also from aircraft, when the shadow of the craft is visible on clouds below.

THE NIGHT INSECTS

Night is a notoriously difficult time to develop an interest in entomology. Not that the bugs are not there: In many places you can count on finding—or, more properly, being found by—abundant numbers of nocturnal mosquitoes. Or you can watch June bugs flying clumsily against window screens and walking on tiptoe across patios and sidewalks. Some mayflies mate after dark, swarming over rivers and lakes in masses so dense their wings hum like electricity.

The easiest way to see night insects is to simply switch on an outside light. They are drawn to it like, well, like moths to a candle. Most species of moths are nocturnal, spending their days camouflaged against the bark of trees or against shaded, protected sides of buildings. Some flowering plants accommodate them by waiting until night to open their blossoms, then emitting fragrances that can attract moths from distances as far as 550 yards. A cruising moth, guided by the sense of smell located in its antennae, seeks out an open blossom and feeds on its nectar with a proboscis it keeps rolled in a spiral when not in use.

There seems little reason a moth should be distracted from its business of sipping nectar and avoiding predators by something as dangerous as an artificial light. Yet, porch lamps and streetlights are irresistible beacons to moths. Moths circle the light as if mesmerized, in frenzied, dizzy orbits. Once caught in their blind orbits, moths become easy prey for bats and nighthawks, or they batter themselves to death, or singe their wings to uselessness on hot surfaces, or die of exhaustion. And nobody knows why.

Male moths have been noticed flying much higher at night than females, behavior that probably helps disperse species and insure genetic vitality. Some entomologists have theorized that the moon could act as a "trigger" to attract male moths, causing them to fly high enough to be scattered across the countryside. If that is the case, attraction to artificial light is a case of clumsy navigation, similar to mating mayflies' tendency to mistake the shining surface of a highway for a river. Moths might be reaching for the light the same way they would reach all the way to the moon if they could.

Another theory is that moths navigate at night by orienting themselves to the light of stars or the moon. Once keyed to a celestial light source, they can travel in a straight line, for instance, by always keeping the light on one side. If by chance they blunder within range of an artificial light they are forced to fly in endless circles because their instinct is to always keep the light on the same side.

As a child my interest in insects was limited mostly to possessing them, and the more difficult they were to catch, the more interesting they became. Few were as elusive or fascinating as the fireflies or lightning bugs—we used the names interchangeably—that blinked on humid summer nights in the field behind our house and along the edge of the woods across the road. My brother and I chased them tirelessly, but they flashed only at intervals and for such short durations that we were seldom able to intercept them in flight. We had better luck capturing those (the females, I now realize) that blinked stationary in the grass or from perches on low branches. We would gather a few and place them in jars. When their glowing dimmed we lost interest, released them, and went searching for brighter prey.

Fireflies and lightning bugs are neither flies nor bugs—they are beetles, members of the family Lampyridae (from the Greek for "shining fire"), which contains some 2,000 species of luminous and nonluminous insects throughout the world. In Europe the common glowworm is a wingless female that shines with a steady light to attract nonluminescent flying males. Both the male and female of most North American species are luminescent, with a luminescent larva that is commonly known as a glowworm. In many species even the eggs glow.

The firefly's fire is produced by an organ at the end of its abdomen that at night can emit a bright green, yellow, blue, or, in at least one species, red glow. Speculation that the light might be a defense used to frighten away predators is probably ungrounded, since the lights could as easily attract as repel. Frogs at least are not put off by it: They have been seen glowing with interior light in the moments after swallowing fireflies. The light probably serves no other purpose than to attract mates. Why such an unusual means of courtship should have evolved in the first place is a matter of conjecture, but some biologists have suggested that it may have originated as a mechanism for burning off excess oxygen at a time when oxygen was new in the atmosphere and may have been toxic to many organisms.

The light organ of a firefly contains cells storing a compound known as luciferin. Surrounding those cells are tubes that can be opened to allow air to enter. When oxygen in the tubes comes in contact with the luciferin, an enzyme called luciferase (both named for Lucifer, the rebellious archangel who was a bearer of light before his expulsion from heaven) is released, which acts as a catalyst and produces the "cold" light characteristic of bioluminescence. It is one of the most efficient light sources known, converting almost 100 percent of its energy to light. An incandescent lightbulb, in contrast, converts 90 percent of its energy into heat and only 10 percent into light. Reflector cells located behind the light organs in the firefly's abdomen direct the light outward. The characteristic on-off flash is controlled by the insect's regulation of its oxygen supply.

Each species of firefly has a specific flash pattern that helps identify them on nights when many different fireflies are active. Males generally fly around, giving off patterns of flashes unique to their species, patterns that vary by color, duration, interval, number of flashes, and distance flown between signals. When a female perched on the ground or on a low branch spots a signal from a passing male of her own species she reciprocates with the same signal, which attracts the male to her. If the chemistry is right, they mate. The female then lays her fertilized eggs on or just beneath the soil. But romance is not without its hazards. The female of the highly carnivorous genus Photuris has turned the lovesick behavior of male fireflies to her advantage. She waits on a branch, watching for their signals. When she spots one she mimics his code, and if he comes to her he is promptly devoured.

Fireflies begin their lives as larvae that feed on snails, slugs, earthworms, and immature insects, which they inject with a toxin that paralyzes the prey and turns their insides to liquid that can be sucked out at leisure. In late fall the larvae burrow underground to spend the winter, emerging in the spring to continue feeding. By early summer they construct a small shell of dirt in which to pupate, emerging two weeks later as a winged adult. The adults of some species are carnivorous, feeding primarily on small insects, while others do not feed at all and are devoted solely to using their flashing lights to attract mates.

Fireflies are uncommon in the Pacific coastal states, but are abundant elsewhere in North America, Europe, Asia, and in most tropical regions of the world. In eastern North America they can often be found near marshlands or over meadows and along the edges of woods.

One of the strangest of firefly displays is enacted by the Pteroptyx fireflies of Southeast Asia, when after darkness falls swarms of the males rise up from the underbrush and gather in trees. At first the thousands of flashes they emit are unusual only in their number, but within minutes, as if on command, the insects begin blinking their lights in unison. Their performance is so nearly synchronous that it can not be explained by each insect reacting to the flash of a neighbor. Instead, it is as if they blink in time to a shared heartbeat, or as if they were wired to a switch flipped rhythmically on and off. Females in search of partners are attracted from great distances to the brilliantly illuminated trees. The insects' ardor is not cooled even by heavy rain, but the rising of a bright moon extinguishes the lights at once.

The brightest of fireflies is probably the tropical American click beetle of the genus Pyrophorus, often called a fire beetle. Measuring more than two inches long, it is equipped with a pair of green lights so bright it is possible to read a newspaper at night by them. Fire beetles are sometimes used as ornaments, tied with a thread and fastened in hair or on clothing, by natives of the tropics. They are also frequently captured and kept in cages where they serve as living lamps.

One of the earliest references to fireflies is in the Chinese *Shih Ching*, or "Book of Odes," which dates from 1,500 to 1,000 B.C., and mentions the "fitful light of glowworms." Aristotle gave some attention to fireflies and glowworms and was sufficiently interested in bioluminescence to note that "some things though they are not in their nature fire nor any species of fire, yet seem to produce light." Pliny explained erroneously that the firefly opened and

closed its wings to produce a blinking light. In Japan, where for centuries captive fireflies have been used to illuminate lamps, it was believed that fireflies were engendered from decaying grass and glowworms arose from bamboo roots.

Like most animals with mysterious qualities, the firefly and glowworm have inspired many mythological explanations for their origin. In India it is claimed that sparks from a fire are transformed into fireflies, and that they are frequently gathered by bats to light the otherwise dark caves where they raise their offspring. Tribespeople of Orissa in India believe that fireflies are the eyes of gods killed in long-ago battles. In Indonesia, a person who becomes seriously ill might be treated by a sorcerer who places a firefly on the patient's forehead to replace the one the soul is believed to have used to escape the sick body.

The elusive blinking lights of fireflies seem to give them more in common with fairies than with insects. It is hardly surprising that Tinkerbell in *Peter Pan* has a distinctly firefly-like appearance. In Woody Allen's *A Midsummer's Night Sex Comedy*, fireflies represent the glowing spirits of men and women who have died in the act of lovemaking. As Jose Ferrer says, just before his character, Leopold, joins the fireflies weaving luminous trails in the darkness around him, "These woods are enchanted."

THE AIRBORNE BESTIARY

Pity my poor neighbor. He is a warrior in the battle against weeds, a Saturday-morning inspector of bluegrass and a Monday-night spreader of Weed Be Gone. He patrols the borders of his yard, head bent, weed-digger in hand, ready to pounce on any intruders. "Look!" he yells to me, holding up the uprooted enemy, "dandelion." There is a slight accusation in his voice. He blames me for his troubles because I happen to enjoy the dandelions thriving in my yard and do nothing to discourage them. In blossom they are scattered across my lawn like constellations in a night sky. My kids like to pick them between their bare toes or rub yellow on their cheeks and say it is butter.

I wander over to watch him engage the enemy, and see, drifting with the breeze like an upright mayfly riding on a river, the white and delicate parachute of a dandelion seed. My neighbor stands abruptly, roots dangling from his hand, and unwittingly intercepts the drifting seed with his hair. It sticks there, perky as a fresh daisy, then catches a breath of air and floats past his shoulder and settles among the grass on that rich and pampered soil. It will have no trouble competing down there.

The wind is a tremendous distributor of life on our planet, and plants, always quick to capitalize on a good thing, have evolved many mechanisms to take advantage of it. Dandelion seeds, riding patiently beneath their umbrella of down, can ride a breeze for hours. The tiny plumed seeds of bulrushes and cattails have been carried hundreds of miles over open ocean to colonize remote islands. A sample of air collected 5,000 feet above the ground over Louisiana contained the seeds of daisies, cottonwood trees, five species of grass, and four species of thistle.

That same sample found more than just plant seeds. Most of us,

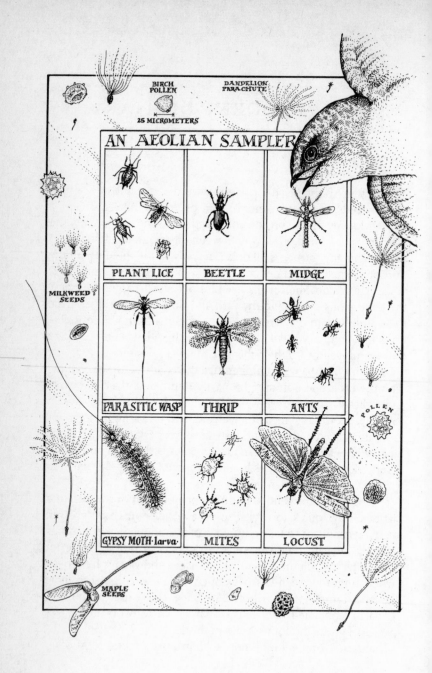

BIRCH POLLEN

DANDELION PARACHUTE

25 MICROMETERS

AN AEOLIAN SAMPLER

PLANT LICE	BEETLE	MIDGE
PARASITIC WASP	THRIP	ANTS
GYPSY MOTH · larva ·	MITES	LOCUST

MILKWEED SEEDS

POLLEN

MAPLE SEEDS

when we think of life aloft, think of flocking blackbirds and migrating monarch butterflies, of mayflies and swallows, of hawks spiraling upward on rising thermals of air. The sky is crossed now and then by interesting creatures with wings, but we think of it as a mostly empty place.

Yet, in summer any cubic mile of sky over the rich temperate regions of the planet contains as many as 25 million assorted insects, spiders, and other animals suspended or drifting in the air the way plankton drifts through the ocean. Countless tiny organisms fill the sky to an amazing height, traveling vast distances on the wind.

High on the ice fields of Mount Everest, at a height of 22,000 feet, lives a species of jumping spider that is probably the highest permanent inhabitant of the earth. Biologists early in the century were baffled by the spider, because the harsh environment where it lived seemed to offer nothing for it to prey upon. They eventually learned, to their astonishment, that the spider had only to wait for its meals to be delivered to it. Every day, countless flies, aphids, butterflies, moths, beetles, ants, gnats, midges, and mites, none of them indigenous to high altitudes, were swept by updrafts to the top of the mountain and deposited on the ice and snow. Since then researchers at high altitude have observed hundreds of insects falling on a small area in a matter of minutes, both in the Himalayas and in the Sierra Nevadas in California.

In the 1930s, a pair of entomologists in England sent up a box kite equipped with a specimen net and in 125 hours of collecting, at heights from 150 to 2,000 feet, gathered 839 insects. The insects included plant lice, small flies, aphids, thrips, and parasitic wasps —mostly small and light-bodied insects with poor flight abilities. About that same time, an entomologist in the United States named Perry Glick flew some 1,450 flights in a biplane equipped with screens between the wings. In the sky over Louisiana he collected 30,033 individual insects representing 700 species, at altitudes from 20 feet to 15,000 feet, and concluded that a column of air one mile square, extending from 50 feet to 14,000 feet above the ground, contained an average of 25,000,000 insects, plus uncountable numbers of seeds, spores, pollens, bacteria, and other minute living things. In 1963 biologist L. W. Swan named this airborne bestiary the "aeolian zone," for Aeolus, the Greek god of the wind.

Not all the voyagers in the aeolian zone are tiny. High alpine plateaus in the Rocky Mountains are often visited by grasshoppers and locusts that have been carried from the valleys below by winds rising up the flanks of the mountain. When the wind lets them down finally they often land on snow or ice, where they die, freeze, and are buried. Thousands of them have ended up embedded in a wall of ice in the Beartooth Mountains in southern Montana, where they lie in strata as deep as sixty feet and as old as hundreds of years. When the ice melts during hot summers, layers of grasshoppers are exposed to the sun, where they provide a freshly thawed meal for birds and bears.

Although most of the animals carried aloft by winds are probably unwilling travelers, there are many species of spiders that have adapted splendidly to air transportation and use it to disperse their young to widely separated parts of the world. On sunny, windy days a spiderling ready to make its own way in the world leaves its mother, who has protected it while it lives off a slowly digested embryonic yolk in its digestive system, and climbs to the tip of a twig, or blade of grass, or a flagpole, raises its abdomen into the air and spins thread of fine silk. It spins and spins, releasing silk until it waves six to ten feet in the air and is caught by the wind. The spiderling then releases its hold on its perch and climbs out onto its thread and is carried away. The journey can be exceedingly long and lofty. Spiders on their gossamer threads have been captured five miles above seal level, and have been reported seen in the stratosphere, more than seven miles high. They have come down in the rigging of ships hundreds of miles from the nearest land.

The adults of some species never lose their ability to sail. Some ground-dwelling jumping spiders, if threatened by a predator on a windy day, will throw up their abdomens, spin off a length of silk, and make their escape on the wind.

When a spider comes down from its travels it releases the thread, which is carried off by the wind. The threads can sometimes be seen, shimmering in the sunlight, as they drift overhead. They often become tangled with other threads, then with others, eventually gathering into sheets of filmy silver gauze too heavy to stay airborne. Certain places, like California's Yosemite Valley, collect large masses of the silk as it is funneled in by the wind and covers trees and rocks until they look, to borrow Jean Craighead George's phrase from *Beastly Inventions*, "like sheet-covered furniture in a deserted house."

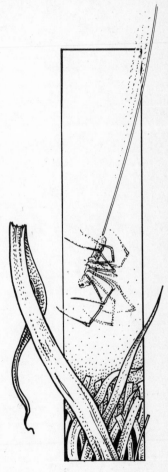

The caterpillar of the gypsy moth is another enterprising aeronaut that makes good use of the wind. This voracious devourer of oak, apple, beech, and birch leaves is a true gypsy, sending out silken strands that catch air currents and carry it away to destinations unknown. The hairs on the caterpillar's body are hollow, increasing its buoyancy and allowing it to be carried as high as 2,000 feet above the ground, and across miles of countryside. Since its first North American appearance in Medford, Massachusetts, in 1868, where it escaped following an amateur entomologist's bungled attempt to raise the species for silk production, the gypsy moth has defeated all attempts at quarantine and has so far spread as far north as Maine and as far west as Minnesota and Texas.

Lying on my back, in the yard, in a circle of dandelion

blossoms, I can look up any summer afternoon and see insects drifting by on the wind. I can't tell how far up they go, but sometimes I can see the darting flight of swallows so high they look hardly larger than insects themselves. Life swirls and eddies to the very limits of inhabitable earth. If the wind could blow across the vast and airless space between the planets, no doubt it would populate the universe.

BIRD AND INSECT WEATHER FORECASTERS

When Mark Twain complained that everybody talked about the weather but nobody did anything about it, he neglected to give credit to the countless people who have set out to do something about it by praying, dancing, shooting cannons in the air, stepping on spiders, and decapitating snakes. If such tactics disappoint us, we settle for predictions. Maybe we can't change the weather, but at least we can be forewarned.

If science fails to accurately predict the weather—and it fails 30 to 50 percent of the time—a lot of us fall back on folklore. At various times and in various places, people have believed these to be reliable signs of impending rain: when dry springs begin to flow, burning wood pops, flowers smell sweeter and manure smells worse than usual, stones begin to sweat, doors and windows stick, dandelion blossoms close, dead branches fall in calm weather, lamp wicks crackle, candles sputter, or rings appear around the sun or moon.

Almost every living thing has been credited with finer sensitivity to approaching weather than we dull, dull humans possess. Our joints and bunions might ache before a storm, due to the buildup of gases in our tissues as the barometer falls, but otherwise we tend to be singularly oblivious to signs other animals find unmistakable. Thus, at least according to the folklore of weather, you can count on a rainstorm coming soon if crickets chirp on the hearth, roosters crow in the evening, owls hoot, frogs croak, goats snort, jackasses bray, doves coo, and flies bite. Likewise, it is certain that rain will soon fall if bees stay near their nests, starlings flock, fish leap, ants travel in straight lines, pigs prepare beds of straw or scratch

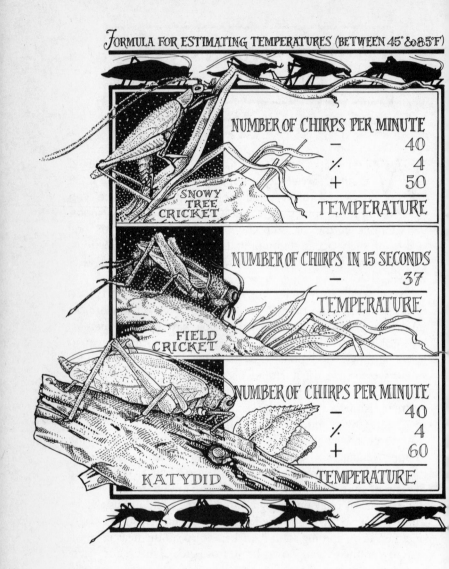

SNOWY TREE CRICKET

NUMBER OF CHIRPS PER MINUTE	
−	40
÷	4
+	50
TEMPERATURE	

FIELD CRICKET

NUMBER OF CHIRPS IN 15 SECONDS	
−	37
TEMPERATURE	

KATYDID

NUMBER OF CHIRPS PER MINUTE	
−	40
÷	4
+	60
TEMPERATURE	

themselves on posts, sheep are frisky, cats cover their ears, dogs roll on their backs or howl or eat grass, horses paw the earth or roll on the ground, cows huddle or scratch their ears or lift their tails, ducks oil their feathers, or spiders fall from their webs. For centuries it was believed that a leech held captive in a bottle acts as an organic barometer: When it climbs high in the neck of the bottle the barometer is falling and rain is imminent; when it lies coiled on the bottom the barometer is rising and you can expect fine weather; when it moves about rapidly, high winds are coming.

Some of the most widespread of weather legends involve birds and insects. Gnats swarming in the open indicates clear weather; swarms in the shade means rain. New England fishermen deciding whether to put to sea in the morning traditionally looked to the gulls: If the birds stayed on the beach, so did the fishermen. Other weather prophets claimed songbirds stopped singing a day or two before a serious storm. Japanese weather-watchers believe that if jays disappear during the first two weeks of September, the height of typhoon season, it is a warning that a storm will appear. The Japanese also listen for the strident chirping of crickets to warn them of the approach of winter.

Some folk observations have proven to have basis in fact. There is evidence to suggest that birds are especially sensitive to low-frequency sound waves emitted from thunderstorms, which might explain the often-noted weather superstition that crows, ducks, geese, owls, woodpeckers, and other birds become especially vocal shortly before stormy weather. The notion that robins stay near their nests before a storm appears in this ancient weather saying:

> If the robin sings in the bush,
> Then the weather will be coarse.
> If the robin sings on the barn,
> Then the weather will be warm.

Such behavior makes evolutionary sense since a robin's nest is frail enough to be destroyed by strong wind and heavy rain, and it is definitely in the species' best interest for the adult bird to protect the nest by sheltering it with its body during a storm. Also, the low-pressure air that comes ahead of a storm is less dense than high-pressure air, and more difficult to fly in. The high pressure and consistent convection currents of fair weather make it easier for birds to climb to high altitudes, thus the ditty:

> *Swallows high in the air*
> *Means the weather will be fair.*

Or this variation:

> *Goose honks high, weather fair.*
> *Goose honks low, weather foul.*

Geese and other large birds often choose to fly at the altitude with the densest air because it offers their wings the most lift. During good weather, high pressure causes that ideal height to rise hundreds or thousands of feet into the air. But when the low-pressure air mass that precedes stormy weather moves in, the best height for flying is low to the ground.

John J. Rowlands, in *Cache Lake Country*, his chronicle of a year spent living off the land in the wilderness of Ontario, reports matter-of-factly that the snowy tree cricket is sometimes called the "temperature cricket" because of its uncanny ability to broadcast the outside temperature: "Just count the number of chirps per minute, subtract forty, divide the result by four, add fifty, and the result will be the temperature within a degree or two. If you don't believe me, try it yourself and you will be convinced."

Rowlands was repeating a formula devised by physicist A. E. Dolbear and published as Dolbear's Law in an article published in 1897 titled "The Cricket as a Thermometer." Although Dolbear failed to specify the type of cricket his formula was derived from, later commentators concluded that the snowy tree cricket was the most reliable of the songsters. The common field cricket is fairly reliable too, but is has a tendency to change its rate of chirps according to its age, success in finding a mate, and other factors besides temperature. Albert Lee, in his book *Weather Wisdom: Facts and Folklore of Weather Forecasting*, reports a slightly altered formula said to apply to field crickets, which he says are called "the poor man's thermometer" because they can be used to calculate the *exact* temperature, more accurately than a mercury thermometer: "Count the number of chirps the cricket makes during a 15-second interval, then add 37 to the number to get the correct temperature in degrees Fahrenheit. If he chirps 40 times in 15 seconds, the temperature is precisely 77 degrees where the cricket is sitting. And it never varies." The same number of chirps from a snowy tree cricket, by Dolbear's calculation, would indicate a temperature of 70 degrees.

Others have claimed the outside temperature can be determined by counting the number of chirps per minute of the katydid, subtracting 40, dividing by 4, and adding 60.

Keep in mind that temperatures tend to be a little cooler down in the grass where crickets and katydids hang out. Also, though it comes on good authority that the cricket method of temperature gauging is uncannily accurate, crickets tend to get sluggish and silent at temperatures below about 45 degrees Fahrenheit, and sullen and silent at temperatures about 85 degrees Fahrenheit.

Animals can sometimes give clues to short-range weather changes, but longer range predictions are a bit trickier. Every year

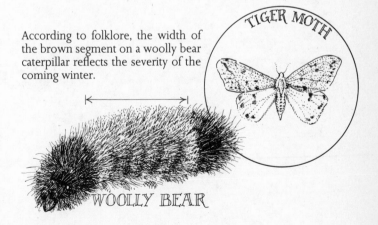

According to folklore, the width of the brown segment on a woolly bear caterpillar reflects the severity of the coming winter.

in New England and the Midwest the newspapers report prognostications of the severity of the winter to come based on the appearance of the woolly bear caterpillar, larva of the tiger moth. According to folklore, if the brown segment on the brown-and-black banded caterpillar is unusually narrow the winter will be harder than usual; if the band is very wide it indicates a mild winter to come. At least one biologist took the claim seriously enough to devote several years to recording band widths and comparing them to weather records. The results of his research seemed encouraging, with some of the data supporting the folklorists' contention, until he discovered that at the same time one population of caterpillars was predicting a tough winter to come, another population living a short distance away predicted an easy winter.

We tend to remember our successes and forget our failures, and we naturally recall the oddest and most unusual things that happen to us. But I can say with absolute confidence that when the ordinarily inoffensive sand flies living along the shore of Lake Michigan begin to gnaw on the exposed ankles and wrists of humans, it means we are sure to have rain within a few hours. It never fails.

Autumn

Autumnal Equinox

Autumn begins with a subtle change in the light, with skies a deeper blue, and nights that become suddenly clear and chilled. The season comes full with the first frost, the disappearance of migrant birds, and the harvesting of the season's last crops. It can begin as early as mid-August, with a day that seems somehow less bright than the one before, with light that is indirect, glancing, beaming at a lower angle from the sky. You notice a new vitality, a stirring as if from slumber. Then a few days later you see the first red leaf in a grove of maples, bright as a drop of liquid ruby.

For many of us in North America, the pure essence of autumn is audible, coming on certain nights in September, October, or November, the first nights cold enough to freeze the abandoned squash in the vegetable garden and leave a skin of ice on ponds and puddles. Without expecting it, without knowing at first what it is, we hear the strange, doglike honking of Canada geese flying south toward warmer weather. They travel at night, arranged in their familiar V-formations, after spending the day feeding in grain fields and meadows. We hear that unmistakable "ha-runk, ha-runk, ha-runk," and something deep and elemental is stirred. It is a wild sound, reminding us that the season has changed, irrevocably, and that at some level we ourselves are connected with wildness.

In agricultural societies autumn has always been the season of harvest and a time of feasting. Harvest traditions still celebrated today include the Jewish holiday Succoth, celebrated each autumn according to guidelines set forth in Deuteronomy: "You shall keep the feast of booths seven days, when you make your ingathering from your threshing floor and your wine press; you shall rejoice in your feast." Another is the English tradition of Harvest Home, which originated as a feast provided by a farmer on the last day of the harvest for the reapers who helped bring in the crops. Memories of that tradition came to America with the Pilgrims, who after their first difficult year in Plymouth Colony in 1621, gave thanks for the harvest with a bountiful feast. The exact date of that first Thanksgiving has never been established. Canadians celebrate it in October; in the United States it was officially decreed for the last Thursday in November by Abraham Lincoln in 1864. Our modern version of Thanksgiving, complete with football games on television, is oddly reminiscent of the traditional harvest celebration of the Natchez Indians, who played ball games before their feast. Going back further, the harvest feast of the ancient Maya included turkey on the menu, and was accompanied by ritual ball games.

"Fall of the leaf," as the season was named in print as far back as 1545, officially begins September 22 or 23, the autumnal equinox, when the noon sun is straight above the equator and day and night are again of equal duration everywhere on earth. By the calendar, it lasts until the winter solstice on December 21. Most of us who live in the northern states think of it, not in calendar pages, but in terms of frost and snow: Fall begins with the first frost and ends with the first lasting snow.

Those months of progressively colder weather are often relieved by a week or two of "Indian summer," that stretch of warm, hazy weather in October, when south winds blow up from the Gulf of Mexico to the central and eastern United States. First used in reference to weather in western Pennsylvania in the eighteenth century, it may have referred to the Indian tradition of changing camps in the fall and moving to winter hunting grounds, or to the strategy of attacking settlers when they were relaxed during good weather. It might have referred also to the typically hazy atmospheric conditions of the season, and an association with fires built by Indians to smoke meat in preparation for winter and to burn brush to prepare for the following year's corn plantings. Or, the term may have originated in the myths of New England tribes that credited the two or three weeks of lovely weather every fall to benevolent gods. It was widely believed that cold weather and storms always occur around the autumnal equinox. It was not until after these brief "half winters" or "squaw winters" that Indian summer could be expected to occur.

Europe rarely experiences an interval of pleasant autumn weather comparable to North America's Indian summer, although warm spells in fall are sometimes called "old woman's summer" or "fool's summer." A more reliable approximation to Indian summer is found in Greece, where the tranquil and warm halcyon days are a regular occurrence in early December. The term refers to Halcyone of Greek mythology, the daughter of Aeolus, who threw herself into the sea in despair when she learned her husband, Ceyx, had drowned. The gods changed her and her husband into kingfishers, or halcyons, and bade them breed on the sea at the time of the winter solstice. To aid them, Zeus forbade the winds to blow for seven days before and after the solstice.

Autumn Leaves

IN MUCH OF THE world it is not autumn without autumn leaves. When green forests grow spotted with yellow, orange, and red, we know it is time to gather the crops, cut firewood, and put up the storm windows.

Contrary to the old legend, Jack Frost does not paint leaves with bright colors. Temperature and frost have nothing to do with it at all. The bright colors of autumn leaves and the subsequent drop of those leaves are a precautionary tactic used by trees to protect themselves during the rigors of the winter season. It would be difficult in winter for deciduous trees to absorb water as quickly through their roots as it is desiccated from their leaves, so they simply refuse to make the effort. As the amount of sunlight diminishes at the onset of autumn, the trees shut down the activity in their leaves, drawing in the sugar and protein stored there and ceasing to produce chlorophyll. The colors that remain in the leaves are carotenoid pigments that were there all along, but were masked in warmer seasons by the bright green of the chlorophyll cells.

Leaf fall begins when hormones stimulate a layer of cells at the base of each leaf stalk to die and form a seal between the leaf and its branch. As those cells congregate and die, they form a corky layer of dead tissue. When the seal is complete it takes only a light breeze to separate the leaf from the tree and send it spiraling to the ground.

Above Autumn Leaves: Migrating Birds

I once listened to a man argue vigorously that ruby-throated hummingbirds manage their long migration to Central America every fall by hitchhiking on the backs of Canada geese. How else, he reasoned, could such a tiny, fragile creature survive a flight of hundreds of miles across open ocean?

The migration of birds each fall and spring has long been among the mysteries of the natural world. Airborne hitchhiking is just one of many theories that have been suggested over the centuries. Aristotle, who was among the first naturalists to propose that birds were long-distance travelers, also believed swallows and some other species hibernated in holes in the ground each winter. Later observers speculated that winter birds burrowed beneath mud and water, a belief common until well into the twentieth century. One of the strangest possibilities was put forth by the Bishop of Hereford in the early seventeenth century in a tract titled *The Man in the Moone, A Discourse of a Voyage Thither*. The good bishop argued it should be possible to harness a vehicle with birds and use it for space travel, since it was widely known that birds migrated to the moon each winter.

A good example of human confusion about the migration of birds can be seen in this confident passage from Pliny's *Natural History*:

> It is a vast distance, if one calculates it, over which they [cranes] come from the eastern sea. They agree together when they start, and they fly high so as to see their route in front of them; they choose a leader to follow, and have some of their number stationed in turns at the end of the line to shout orders and keep

the flock together with their cries. At night time they have sen-
tries who hold a stone in their claws, which if drowsiness makes
them drop it falls and convicts them of slackness. . . . It is certain
that when they are going to fly across the Black Sea they first of
all make for the straits between the two promontories of Rams-
brow and Carambis, and proceed to ballast themselves with
sand; and that when they have crossed the middle of the sea they
throw away the pebbles out of their claws and, when they have
reached the mainland, the sand out of their throats as well.

Until fairly recently the migration of the world's hummingbirds
remained a hotly debated controversy. The major argument centered around fuel capacity. Hummingbirds use such vast amounts
of energy during flight—beating their wings fifty to seventy-five
times per second—that it is necessary for them to eat almost continuously to survive. Some species consume their own body
weight in nectar and insects each day, seldom resting more than
fifteen minutes between meals. To achieve comparable calorie intake, an adult human would need to eat 285 pounds of hamburger
each day. Because of such strict dietary requirements, hummingbirds tend to migrate by buzzing from flowerbed to flowerbed,
literally eating their way north and south each year.

In the case of the ruby-throat, the route followed during that
migration creates a dilemma. Ruby-throats, the only North American hummingbird found east of the Mississippi, prefer to spend
their winters in Central America. Their migration takes them to the
shores of the Gulf of Mexico, where they launch a 500- to 650-mile
journey across open sea—a journey that scientists once believed
was impossible for the tiny birds. How could they survive such a
strenuous flight, one that required a minimum of twenty or
twenty-five hours of top-speed effort, over open ocean? With their
extraordinary rates of metabolism demanding almost constant re-
fueling with nectar and insects, surely the birds would die before
reaching land.

Yet, the survival rate of the birds using the transoceanic route is
apparently very high. Scientists scratched their heads. Laymen decided the birds must catch rides on sturdier fliers.

Researchers eventually discovered that ruby-throats utilize a simple tactic to survive the open-sea crossing. In just a few weeks of heavy eating each bird is capable of increasing its body weight by half, until up to 40 percent of its bulk is composed of fat. With that much stored energy in reserve, the 500-mile crossing of the Gulf of Mexico, at an airspeed of thirty miles per hour, becomes fairly simple.

But why bother migrating? Climate and food are frequently cited reasons, but most birds migrate much farther than necessary just to find abundant food and favorable weather. Others leave long before the weather turns cold and the food scarce, and some species do not migrate at all.

One argument is that today's migrations might be a throw-back to the ice age of 11,000 years ago, when harsh climate over much of the planet must have severely affected bird life. Another theory is that the Southern Hemisphere may have been the original home of many species of birds that now live in the northern portions of the planet and that their annual migrations are simply a return to ancestral territory.

Regardless of the reasons, it is indisputable that migrations, although they may cost certain species of birds half their total population in casualties each year, have logical advantages. The northern temperate regions, where the majority of migratory birds nest, are far more spacious than those lands in the Southern Hemisphere. A full three-quarters of the northern region is land, while 90 percent of the southern zone is taken up by ocean. With more land in the north there are more opportunities for feeding growing broods of young. Also, and just as importantly, the longer summer days of the northern regions are critical. At the equator, there are twelve hours of daylight on June 21; at the 40th parallel the day has lengthened to fifteen hours; and at the 60th parallel, the northern Canadian regions where so many of the continent's birds nest, it extends to nineteen hours—long enough to put in the full day necessary to feed rapacious nestlings.

And the Prize for the
Longest Migration Goes to . . .

A MOST AMBITIOUS MIGRATION is undertaken twice each year by the bristle-thighed curlew, a small shorebird that nests in the islands of the Bering Strait and migrates to remote Pacific Islands each winter. From Alaska, it flies south across 2,500 miles of open Pacific Ocean to the nearest landfall, Hawaii. After a brief rest it sometimes chooses to travel another 2,100 miles south to the Marquesas Islands.

The bobolink has one of the largest migration routes of any of the passerines, or perching birds, traveling more than 6,000 miles from its breeding grounds in the northcentral United States and southern Canada across the Caribbean to South America, wintering as far south as the pampas of Argentina.

An even smaller bird with an impressive migration is the black-poll warbler. Breeding in the northern United States, it spends winters in South America after taking off from the New England coast of North America and crossing more than 2,300 miles of open water in the Atlantic Ocean and Caribbean Sea. The open-water crossing takes an average of eighty-six hours of nonstop flight and depends so much on favorable winds that the tiny birds have been spotted flying at a height of 21,000 feet—nearly four miles above the ocean. Like the ruby-throated hummingbird, the black-poll warbler prepares for its ocean travel by stocking up on food, nearly doubling its weight with layers of fat to sustain it through the journey. Bird banders sometimes refer to the fat deposits, easily felt beneath the skin of the birds, as fuel tanks.

The longest known migration of any bird belongs to the Arctic tern, a small colonial bird that breeds within a few hundred miles of the North Pole. For reasons biologists have yet to fully understand, Arctic terns set out on a journey each spring and fall that take them literally to the ends of the earth, from the Arctic to Antarctica. The trip, furthermore, is circuitous, taking the birds either along the western shores of North and South America, or along the western edges of Europe and Africa. In all, an Arctic tern may travel as many as 25,000 round-trip miles per year.

Complex Navigation Systems

EXACTLY HOW BIRDS navigate so accurately during their migrations has long baffled scientists. Recent research suggests they use a combination of guidance systems, of which the most obvious is visual orientation, to track familiar landmarks such as river valleys, mountains, and ocean shores. They also probably steer by the sun, moon, and stars, keep track of their progress with a highly developed sense of time, see with the aid of polarized and ultraviolet light, hear low-frequency sounds (such as the crashing of surf many miles away), tune into the magnetic field of the earth, and ride with prevailing winds, thanks to a finely tuned sensitivity to approaching weather.

For many years it was assumed that young birds learned migration routes by flying in the company of older birds that had already made the journey. Such guided tours were discredited when young birds that had never migrated were captured, held until the older birds had departed, and were then released. The inexperienced birds were found to have an unerring sense of direction and reached traditional wintering areas without difficulty.

Other landmark studies have demonstrated that birds are fully capable of navigating by the sun, moon, and stars. In one experiment, Dr. Stephen T. Emlen of Cornell University studied the behavior of indigo buntings in a planetarium. Lengthening and shortening the amounts of daylight to correspond to spring and fall migration periods, Emlen found that the birds adjusted their flight to the stars in the artificial sky and flew in directions that corresponded to north and south in the actual sky. Emlen discovered, by progressively blacking out the planetarium's lights, that the birds tended to navigate by the high, reliable constellations in the vicinity of the North Star.

Although such experiments seemed to answer the larger questions about bird navigation, certain implausibilities were not explained. For instance, in 1951 a Manx shearwater was captured from its nesting site on Skokholm Island, Wales, flown by plane across the Atlantic, and released in Boston. It returned to its nest within thirteen days. In other experiments, birds (often homing pigeons) have been kept in light-proof boxes, driven by circuitous routes to far distant places, and released. Most of the released birds settled on the proper direction home almost immediately, within thirty seconds in many cases.

Researchers recently discovered that birds are equipped with a biological compass that orients them to the magnetic field of the earth. Pigeons (as well as bees, some bacteria, and perhaps many other animals), have tiny flecks of magnetite, a compound of iron and oxygen, embedded in their brain tissue. William T. Keeton, of Cornell University, theorized that the magnetic material makes birds aware of the north-south orientation of the earth's magnetic field. He tested his theory with two now-famous experiments. First, he released homing pigeons that had been blindfolded to deprive them of visual orientation. The birds homed in the correct direction. He then attached small magnets to the birds' necks before releasing them. The magnets apparently disrupted the internal compasses, causing the pigeons to become immediately disoriented.

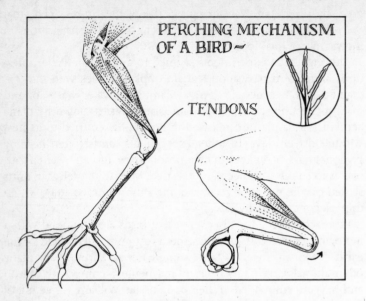

Migrating Birds at Rest:
How Do They Stay on Their Perches?

WHILE RESTING FROM their daily (or nightly) flights, migrating birds alternate feeding with sleeping, gathering energy for the next leg of their journey. Birds that perch accomplish the tricky maneuver of clinging to a tree limb or other roost while asleep, thanks to an interesting anatomical device. From muscles in each thigh a tendon called the flexor tendon extends downward, over the knee, down the length of the shank, around the ankle, and under the toes. The tendons are arranged so that body weight, when bending the knee, causes the tendon to draw tight over the pulley formed by the knee, closing the claws. The resulting grip, in a bird's relaxed state, is so tight that not even strong winds can dislodge the sleeping bird. In fact, birds are sometimes found clinging tenaciously to their perches hours or even days after they have died.

The Tale of a Hibernating Bird

ARISTOTLE WAS partly correct when he proposed that some birds spend the winters in holes. While several species, including some hummingbirds and swifts, enter periods of torpidity, or suspended animation, to preserve energy while they sleep, only one bird is known to actually hibernate as an alternative to migration. The poorwill, *Phalaenoptilis nuttali*, a relative of nighthawks and whip-poorwills, has been found to hibernate for months at a time in rock crevices in the southwestern United States. From a normal temperature of about 104 degrees, the poorwill's body temperature drops to as low as 64 degrees; it then enters a slumber so complete its heartbeat and breathing are nearly undetectable and it can be handled without waking. While it is not known if all poorwills hibernate every winter, the biologist Edmund Jaeger, who documented the first case in Colorado's Chuckwalla Mountains in 1946, found one individual inhabiting the same shallow crevice during four successive winters.

Observing Migrating Birds

IT HAS BEEN estimated that six to seven billion birds can be found in the United States during the breeding season, and that two-thirds of them, or four to five billion birds, migrate each spring and fall. What likelihood is there of observing bird migrations? In some places, and with some species, the likelihood is very great. Some birds, especially shorebirds, waterfowl, and raptors, migrate during the day and are often plainly visible winging north in spring or south in winter. At sites like Hawk Mountain in Pennsylvania, it is possible to see thousands of hawks, eagles, and falcons in a single day as they ride the thermal currents along the mountain. Nighttime bird watchers often choose moonlit nights to take counts of flocking birds as they cross before the moon. Boaters in the Atlantic, Gulf of Mexico, and Great Lakes often find their boats used as temporary rest-stops for exhausted migrants. In the daytime, waterfowl are easily spotted in their distinctive V-formations.

Many birds, however, migrate at night and are difficult to observe. The American woodcock, for instance, travels only after dark, and usually alone or in small, loose groups of scattered individuals. During the day they hide in dense thickets of aspen and alders, resting and feeding on earthworms, and are seldom seen except by hunters.

In autumn, look for the most concentrated migrations during periods of fair weather when the winds are from the north. Migrating birds often make use of winds and weather patterns in their flights. The Arctic tern catches the prevailing northwest winds to fly from Baffin Bay to southern Greenland to western Europe, then uses north and northeastern winds to fly from western Europe to northwest Africa, then easterly trade winds to fly to eastern South America, where it picks up north and northwestern winds to complete its trip to the South Atlantic.

Diligent stocking of bird feeders will attract migrants, even rare and unusual species, especially when nearby shrubs and trees offer cover and protection. Expect migrants to appear in small flocks, although some species, like sparrows, starlings, and blackbirds, can appear suddenly in enormous numbers. Don't count on seeing many hummingbirds in southward flight: They travel at night, alone or in small groups, and seldom or never in the company of larger birds—not even the most accommodating of Canada geese.

MIGRATING INSECTS

The migrations of insects are at least as complex as bird migrations, and in some cases even less understood. Among the best-known migrators are locusts, of biblical fame, which periodically undergo sudden increases in population known as irruptions, and spread across the land in enormous swarms. Other insects that migrate include some species of butterflies, moths, bees, wasps, ladybugs, dragonflies, ants, and termites.

We usually think of butterflies as delicate, even ethereal creatures. Most species live only one to three weeks and spend their entire short lives near the place where they first emerged from their pupae. Yet a few species engage in arduous migrations that are among the most remarkable in the animal kingdom. Because their lives are so short, individual butterflies rarely make the complete round trip during a migration; rather, it is successive generations, born en route, that complete the travels. The Painted Lady, or Thistle butterfly, winters in North Africa and migrates across the Mediterranean in early spring and onward north across Europe, sometimes traveling a thousand miles or more into Scandinavia, northern Russia, and even Iceland. Columbus noted vast flocks of migrating butterflies off the coast of Cuba during his voyages to the New World. Darwin, while traveling off the coast of South America on the HMS *Beagle*, reported seeing enormous clouds of butterflies over the ocean flying in a column he estimated to be 600 feet above the water, a mile wide, and many miles long. He wrote in *The Voyage of the Beagle* that the butterflies appeared "in bands or flocks of countless myriads, extending as far as the eye could range. Even by the aid of a telescope it was not possible to see a space free from butterflies. The seamen cried out 'it was snowing butterflies,' and such in fact was the appearance."

English naturalist and novelist W. H. Hudson witnessed a migration of Vanessa butterflies in a dense three-mile-wide column over the Argentina pampas that lasted for seven or eight hours each day for three days in succession. He estimated the flock contained more than 75 million individual butterflies.

No butterfly migration is as well known and easy to observe as that of the monarch of North America. Yet, only in the last few decades have the migration routes of this familiar butterfly been recognized, and, incredibly, it was not until 1975 that the winter destination of the majority of the migrators was discovered. Prior to that, it was known only that a small percentage of the population moved west each fall, wintered in forests along the California coast between Los Angeles and Monterey, then migrated eastward again in the spring, mating and reproducing along the way. The bulk of the butterflies from the eastern half of the United States apparently migrated south, where some congregated in isolated "butterfly trees" in Florida and other southern states. The exact destination of the majority of the butterflies was not known. They seemed to vanish each fall, causing some scientists to speculate that they simply flew out to the Gulf of Mexico and perished.

Finally, an entomologist named Fred A. Urquhart, who devoted more than forty years to the mystery of monarch migrations, developed a way to identify individual butterflies by tagging their wings with lightweight adhesive strips. Gradually, over a period of years, and with the help of hundreds of volunteers who tagged the monarchs and reported on their appearances, he narrowed the search for their wintering grounds to a region of the Sierra Madre mountains in central Mexico. In January 1975, a friend of Urquhart's, working with directions from the entomologist, stumbled on a twenty-acre area of forest literally alive with wintering monarchs. An estimated 1,000 trees were covered almost entirely with a living membrane of semidormant butterflies. At times their combined weight was enough to break three-inch diameter limbs. It was the mother lode: the winter home of most of North America's monarch butterflies.

Migratory grasshoppers, or locusts, have been responsible for

more destruction of crops than any other creature on earth. When crops are destroyed, famine and death follow, so it is little wonder that infestations of locusts have long been considered a plague, as they were in this excerpt from Exodus: "And the locusts went up over all the land of Egypt . . . very grievous were they . . . for they covered the face of the whole earth, so that the land was darkened; and they did eat every herb of the land, and all the fruit of the trees . . . and there remained not any green thing in the trees or in the herbs of the field through all the land of Egypt."

In some parts of the world magical ceremonies seemed the only way to repel the grievous hordes. Bronze and copper images of locusts were erected in early Naples to rid the city of the insects. In Albania, women formed a funeral procession, carrying a few locusts to a river or stream where they were ritualistically drowned in hopes of frightening away the entire swarm. Villagers in India would capture a single locust, mark its head with a red spot of dye, and release it with the hope that it would lead the other locusts away. Likewise, a single locust would be captured by the Wajagga people in Africa, its legs tied together so it could not walk on the land, and thrown into the air to fly off and lead other locusts to far-off regions. Ethiopians in 1590 reportedly excommunicated a plague of locusts, then had the satisfaction of seeing thousands of them eradicated by a sudden storm.

An infestation of locusts in 1926 in the Sudan and Arabia was described by C. B. Williams in his book *Insect Migration:* "An hour or so later the first outfliers began to appear—gigantic grasshoppers about six inches across the wings, and of a deep purple-brown. Minute by minute the numbers increased, like a brown mist over the tops of the trees. When they settled they changed the color of the forest. . . . The swarm was over a mile wide, over a hundred feet deep, and passed for about nine hours at a speed of about six miles per hour: at two insects per cubic yard (an under-estimate) it must have contained about ten thousand million locusts."

In 1906 multitudes of brown locusts left the Kalahari Desert and decimated the farmlands and veldt of South Africa. One swarm was reported to be twenty miles wide and seventy miles long. The African locust is usually a solitary creature, but in the space of twenty-four to forty-eight hours it can transform into a migratory phase and joining company with millions of others, can destroy millions of acres of cropland. In its solitary phase the insect is docile; researchers found that by simply placing an individual in the solitary phase into a cage with others like it a hormone was stimulated that caused the locusts to shrink in size, darken, and enter the migratory phase. A migrator placed in isolation will likewise revert back to the solitary phase. Once migration commences, however, they gather quickly into swarms containing millions of individuals, each of which requires its own weight in food each day. A single swarm can destroy 80,000 tons of grains and other crops per day.

Mass irruptions of locusts have been a plague not only in Africa and the Middle East. In North America, 1874 became known as the infamous Year of the Locust because of a mass migration of the insects that swept through the Midwest, destroying crops and ruining farmers already shaken by several years of drought. The locusts appeared without warning in July and August that year.

They passed over Lincoln, Nebraska, from nine in the morning until three in the afternoon in an unbroken column estimated to be at least 300 miles wide from east to west and averaging at least half a mile in depth. When that cloud came to earth, the results were disastrous. Written accounts from the period reported almost unbelievable destruction. From *This Place Called Kansas*, by C. C. Howes:

> With a whizzing, whirring sound the grasshoppers came from the northwest and in unbelievable numbers. They lit on everything. I was covered from head to foot . . . at once the insects began to eat. . . . We had about fifteen acres of corn which older settlers said would make fifty bushels an acre. The hoppers landed about four o'clock. By dark there wasn't a stalk of that field corn over a foot high left in the entire field. . . . That night the hoppers ate my straw hat, or most of it, leaving me only a part of the brim and a part of the crown. They seemed to like sweaty things and ate around the sweatband of my hat. They gnawed the handles of pitchforks and other farm tools that had absorbed perspiration and they ate the harness on the horses or hanging in the barn.

From *The Sod-House Frontier*, by Everett Dick: "They came like a driving snow in winter, filling the air, covering the earth, the buildings, the shocks of grain, and everything. . . . Their alighting on the roofs and sides of the houses sounded like a continuous hail storm. They alighted on trees in such great numbers that their weight broke off large limbs. . . . At times the insects were four to six inches deep on the ground and continued to alight for hours."

The locusts that descended on Missouri, Kansas, Nebraska, and much of the Midwest in 1874 were the now apparently extinct Rocky Mountain locust, which was then common to the plateaus and high plains of Colorado, Wyoming, and Montana. Normally the locusts were content to stay in those mountain states, but if their population grew high enough, and drought caused a shortage of the grasses they fed on, they would take to the air and ride with prevailing winds in search of fresh food. Some of the swarms were found to travel from Montana to Texas, a distance of 1,500 miles in seventy-five days, for an average of twenty miles per day.

Considering the precedent of the plague described in Exodus, it was no wonder many people thought the locusts were sent by God to punish them for their sins. When a comet appeared that same summer it seemed, to some, final evidence that they had made a dreadful mistake coming west to this forsaken land. By the end of August hundreds of families had abandoned their farms and were returning east.

Those who remained called on divine assistance. In the spring of 1875, when it seemed likely that the offspring of the previous summer's hoards of locusts would return to finish off the last living things in the Midwest, Missouri's governor issued an official proclamation asking that June 3 be set aside as "a day of fasting and prayer, that the Almighty God may be invoked to remove from our midst those impending calamities, and to grant instead the blessings of abundance and plenty; and the people and all the officers of the State are hereby requested to desist, during that day, from their usual employments, and to assemble at their places of worship for humble and devout prayer." When, a few days after the fasting and prayer of June 3, the locusts began to die or fly off, the governor's action was widely applauded. His reelection was assured, and he found it unnecessary to mention that a few days before issuing his proclamation he had been given an entomologist's detailed report predicting the locusts would leave Missouri by early June.

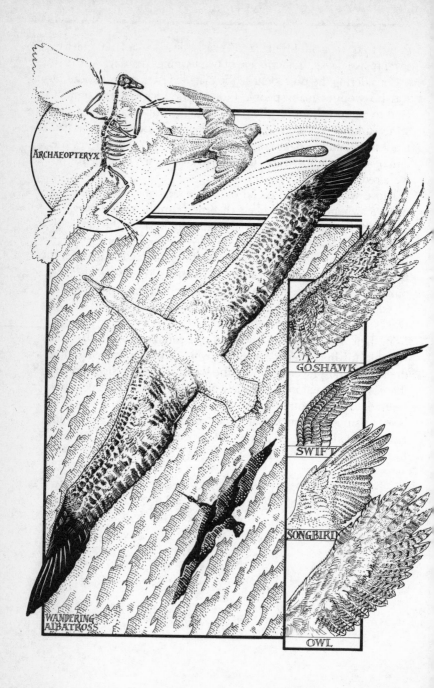

ARCHAEOPTERYX

GOSHAWK

SWIFT

SONGBIRD

WANDERING
ALBATROSS

OWL

BIRDS IN FLIGHT

As impressive as the migrations of birds and insects can be, they are no more impressive than the fact of flight itself. Creatures of the earth have flown since the earliest descendants of the dinosaurs hopped from the ground or launched themselves from trees and ledges and began gliding to escape enemies and search for food. Some flying reptiles took the ability a step further and learned to glide for greater and greater distances. Some, like the pterodons, with their forty-foot wingspans, must have been gliding terrors.

It was apparently not until *Archaeopteryx*, a feathered and probably warm-blooded flier from the Jurassic period, 150 million years ago, that significant progress was made in the evolution of birds. The metabolism of the pigeon-sized *Archaeopteryx* provided it with more power for the strenuous work of flying, but it also made it vulnerable to cold. In time, its reptilian scales evolved into the first feathers.

Every detail of bird anatomy relates to the ability to fly. Birds are equipped with hollow, flexible bones, in a skeleton shaped and reinforced to withstand flight stress. Their bodies are of elegant aerodynamic shape. They have no elaborate—and therefore, heavy —reproductive organs. Their feathers, weighing virtually nothing, insulate them and create aerodynamic lift. In many species, air sacs are found beneath the skin and among the muscles and internal organs. Because flight requires so much energy and a bird's body temperature remains so high, the lungs of a bird are permeated with air chambers that connect to those air sacs throughout the body—an internal air-conditioning system. During flight, air literally passes through the bird's body, cooling it, expelling excess humidity, exchanging carbon dioxide for oxygen. Even a bird as seemingly inelegant as a pelican is a virtual flying wonder: All the bones of a twenty-five-pound adult bird, including beak, skull, and feet, have a combined weight of only twenty-three ounces.

The designers of the first airplanes spent a lot of time trying to understand how a bird's wings worked to lift its body. They eventually discovered the secret was in its distinctive airfoil shape. In cross section, a bird's wing—and an aircraft's wing—is round and blunt at the leading edge, convex on top, concave on bottom, and tapered on the trailing edge. That shape causes air to flow quickly over the top of the wing and to flow slower, with an upward direction on the bottom, creating a lifting action. The wing lifts as long as air flows over and under it.

The lifting of the airfoil shape of wings makes four types of bird flight possible. *Gliding* is performed with the wings stationary and always leads eventually downward since wing lift without thrust is not sufficient to overcome all the force of gravity. If the bird glides over strong currents of rising, heated air—the same currents that produce fair-weather cumulus clouds—it can shift from gliding to *soaring* and rise upward, usually in a spiraling flight, as high as the thermal current reaches. To get airborne, or to increase altitude when there are no thermals to ride, a bird must flap its wings. *Flapping* is a complex physical action with the simple consequence of pushing air down and back to propel the bird up and forward. Feathers on the outer parts of a wing remain flat and airtight on the downstroke, then open slightly to vent air on the upstroke. The fourth type of flight, *hovering*, is a way of remaining stationary in flight, either by rotating the wings in figure-eight patterns from the shoulders, as a hummingbird does, or holding the body in a vertical position, beating the wings rapidly, and opening and closing tail feathers to serve as a baffle, a technique used by kestrels, kingfishers, and many other birds while hunting.

Because different birds have different needs, the shapes and designs of wings vary greatly. Birds like eagles, hawks, and vultures, that glide and soar a great deal in open country, are equipped with broad, long wings that create maximum lift with minimum flapping. Those like falcons and swifts that feed on fast-moving prey in open country require long, pointed wings that can propel the birds at high speed with little drag. Woodland birds like grouse and many songbirds, that must fly quickly for short distances through close confines, have short, rounded wings. Owls, because they hunt in the quiet of the night, have evolved wings with special fringed feathers on the leading edges and softer, looser feathers on the trailing edges that make their flight virtually soundless.

Flying is the most strenuous activity in the animal world, so it makes sense that birds would be interested in fuel efficiency. One of the ways in which they get the most energy output from their calorie intake is to utilize economic styles of flight. Many birds alternate flapping with gliding. The duration of the glide varies from species to species and with the wind strength and direction, but typically a bird will flap one or two beats and glide, repeating the pattern in an undulating flight that is far more efficient than steady flapping.

Most small birds are unable to glide for more than very short distances, so they use a peculiar bounding flight that is probably still more efficient than steady flapping. After a few bursts of vigorous flapping they fold their wings tight to the body to reduce drag and hurtle through the air. They lose altitude while they hurtle, flap rapidly a few times to gain height, then hurtle again. Watch for this distinctive flight pattern in sparrows, finches, warblers, woodpeckers, and many other small birds.

The familiar V-formation of geese and cranes helps them save energy during long migrations. A wing produces an upwash of air spreading out behind it. When a bird enters that upwash, with its wing just behind the tip of the wing of the bird in front, drag is reduced and less power is needed to fly.

Seabirds like skimmers, pelicans, shearwaters, and albatrosses take advantage of an aerodynamic phenomenon known as ground effect to increase the efficiency of their flight. Ground effect occurs whenever a bird—or aircraft—flies at a height above ground or water that is less than the span of its wings. Air passing between the ground and the wing reduces drag and allows easier, more fuel-efficient flight.

The most efficient of all flight styles is soaring, and the king of the soarers is the albatross—the mysterious seabird of Coleridge's "Rime of the Ancient Mariner." The wandering albatross of the

South Atlantic is the largest of living birds, with a wingspan of ten to twelve feet. Young albatrosses leave the subantarctic islands of South Georgia, Prince Edward, and Antipodes, where they are hatched and fledged, and may not touch land again until they are ready to breed two years later. Much of that time aloft is spent soaring.

Because of the difficulties of observing birds in the open seas, relatively little is known about the albatrosses. The English poet Oliver Goldsmith was fascinated by the bird's apparent ability to remain in flight indefinitely and addressed the problem of sleep thus: "At night, when they are pressed by slumber, they rise into the clouds as high as they can; there, putting their head under one wing, they beat the air with the other, and seem to take their ease."

More reliable observers have noted that albatrosses choose to rest on the ocean during calms, then take wing during high winds. Others have observed that the bird is so aerodynamically efficient it can sleep on the wing, not by alternately flapping one wing and resting the other, but by gliding on updrafts thrown up by the dynamics of wave action, which allow the bird to glide for days on end without a single wing-beat. Wandering albatrosses fitted with radio transmitters have been found to wander the ocean in search of food—while leaving a mate waiting patiently on the nest—at speeds up to fifty miles per hour, covering up to 9,320 miles between return visits to the nest.

The secret of an albatross's apparently effortless flight is called "dynamic soaring." Even Goldsmith noted that the birds "alternately ascend and descend at their ease," rising above the sea, gliding down until their wing tips sometimes slice the water, then rising up again—almost never with a flap of wings. The birds achieve perpetual gliding flight in this way because winds that are slowed by friction with waves at water level increase with altitude, creating greater lift on the birds' wings as they rise in height.

Perhaps because of the complex arrangement of feathers and muscles needed in adulthood, a young albatross remains in the nest from nine to twelve months, longer than any other bird. Hatched in the summer, the nestling stays in the nest through the winter, and only begins practicing for flight during the following summer, when with much preliminary stretching, flapping, running, and jumping with wings extended, it finally takes to the air, its true element.

Meals on the Wing

WHEN IT COMES, it comes without warning. The mourning dove, flying across open terrain on its way to its roost, probably senses the onrushing danger but by then it is too late. A peregrine falcon has been trailing from behind, hundreds of feet above it, and now, in a maneuver ornithologists call stooping, it tucks its wings to give its body the aerodynamic profile of a bullet, and plummets. It reaches a top velocity of 125 miles per hour, perhaps as much as 175 miles per hour. At the moment before collision it swings its feet forward, and they strike the dove from behind. The dove folds, killed instantly by the impact, and tumbles in a trail of loose feathers to the ground.

Birds with superb flying skills find the sky a bountiful place. Peregrine falcons and a few other raptors are quick enough to prey on other birds, but the majority of species that feed on the wing feed on insects. The swifts, swallows, and martins are darting, gliding, swooping hunters of the air, adept at switching directions suddenly or turning virtually upside down to snare a flying insect in flight. They spend most of each day aloft, especially during the nesting season, when they must work full time to keep nestlings supplied with insects. The common swift of Europe routinely logs 500 or more miles per day as it flies through the sky gathering insects to feed its young.

Adaptations breed adaptations. When insects turned to the night to escape the birds that preyed on them in daylight, they in turn inspired some birds to begin hunting nocturnally. The nightjars, a large family that includes the whippoorwills and nighthawks, have evolved remarkable flying abilities that enable them to spend more time in the air than almost any bird species. More remarkable than their flight abilities, however, is the way they capture insects on the wing. The nightjars are equipped with short, liplike beaks that can open wide, allowing the birds to fly with their mouths open, sifting insects from the air the way a baleen whale sifts krill from the ocean.

Synchronous Flight

IN THE UPPER MIDWEST, starlings begin flocking early in autumn as they prepare for their migration south. Each evening, birds that have spent the day feeding in fields and meadows and shopping center parking lots rise into the air and fly in broad, sinuous flocks to their roosting areas for the night. As they pass overhead at speeds of twenty-five to thirty-five miles per hour they all switch first one way, then the other, the entire flock as synchronized as a chorus line. Ask any child to observe the flight and you will hear the same question: "How do they all turn at the same time?"

Birds flock for the same reason minnows school: to protect themselves from predators. The sudden, shifting turn of a mass of individuals confuses an attacking predator, causing it to strike blindly at the mass rather than choosing an individual. Those that are captured are usually the weaker stragglers at the fringes of the flock. A few are lost, but the majority are safe.

Still, the question remains: How can a flock of closely arranged birds make those apparently instantaneous switches, dives, swoops, and banking turns without crashing into one another? One long-held theory—that a leader somehow signals its intention to fly left, then right, then left again, directing its obedient flock like a band leader—was disproved when rapid-fire photographs of flocking birds showed them constantly changing leaders. The naturalist Henry Beston watched flocking seabirds and wondered, "Does some current flow through them and between them as they

fly?" That same idea has been elaborated on in an as yet unproved theory that birds, fish, and other animals that gather in aggregates are capable of communicating with bioinformation transfer, a high-speed exchange of information by means of electromagnetic fields. According to that theory, the decision to swoop left is passed through all the birds virtually at the speed of light. If there is a conflict, with some birds sending the message to turn right while others send the message to turn left, the flock splits, perhaps coming together again after a moment, flowing through the air the way running water forges winding, tendril-like routes through sand.

Other research is concentrating on mathematical chaos theory, exploring the idea that flocks of birds behave in ways that can be duplicated with a computer programmed to follow a set of pre-scribed rules. Frank H. Heppner, a zoologist at the University of Rhode Island who has studied the mystery of flocking synchron-ism for more than twenty years, proposed that four rules can be applied to birds in flight: 1) Birds are attracted to a focal point, such as a roost, and the closer they get to it the stronger its attraction; 2) birds are attracted to one another (because there is safety in num-bers) but if they get too close, they are repelled to avoid collisions; 3) birds want to keep a steady velocity; 4) the path of a bird's flight can be altered by random factors such as gusts of wind or the sudden shadow of a hawk. When Heppner programmed all those rules into a computer and animated a collection of figures on the screen to represent a flock of birds, their motion closely duplicated the actual flight behavior of flocking birds.

The research, though yet inconclusive, suggests that birds of a feather are motivated in unison by the same desires and fears. Put those birds together in a flock, in flight, and they perform in beau-tiful, furious concert.

Insects in Flight

In my dreams of flight I have the grace of a chimney swift and the speed of a peregrine falcon, but if I really wanted aerial excellence I would wish for the skills of a housefly. Insects are the best flyers on earth, having had hundreds of millions of years, far longer than birds, to perfect their techniques and evolve specialized anatomical features. The wings of insects, rather than being modified limbs as they are in all other flyers, are designed strictly for flight. Covered with chitin, the same light, resilient material that composes the external skeletons of insects, strengthened by tubular support veins, and manipulated by powerful chest muscles, their wings are incredibly efficient tools. The fastest known physical action of any living thing is the beating of a common midge's wings. They beat at a normal rate of 57,000 times per minute—950 times each second, compared to the 250 beats per second of a honeybee or the 75 beats per second of a ruby-throated hummingbird—but in an emergency a midge can increase the rate to more than 2,000 beats per second.

Insects were already abundant when the first amphibians appeared in the world. Flight made them safe from insect-eating dinosaurs, until those dinosaurs learned to leap and glide and eventually evolved into fliers themselves. The fossil evidence suggests that the first birds may have annihilated in a very short time entire species of slow-flying, otherwise-defenseless insects. To protect themselves, insects evolved other strategies like camouflage, minute size, nocturnal feeding, and swift, evasive flight.

Evasive flight proved very successful for one of the oldest of flying insects, the dragonfly. Little changed since their ancestors hunted in jungles 250 million years ago, the 4,700 species of dragonflies in the world today remain awesome predators of small

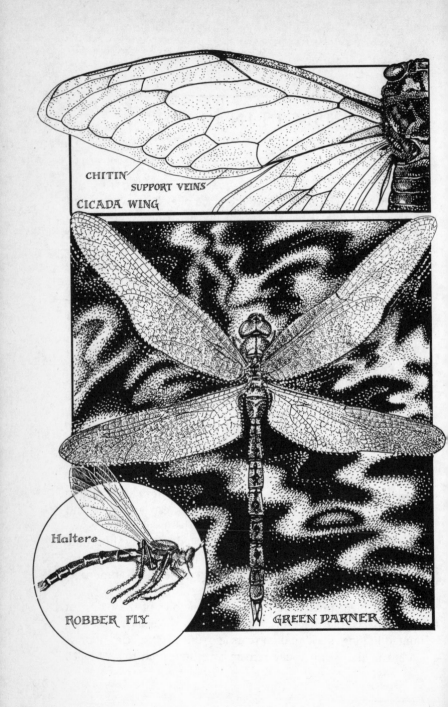

CHITIN
SUPPORT VEINS
CICADA WING

Haltere
ROBBER FLY

GREEN DARNER

flying insects and skilled evaders of predators themselves. Ento-mologists consider dragonflies *palaeopterous* (or "ancient-winged") because, unlike more recently evolved flying insects, they have no way to fold their two sets of wings backward to lie against their bodies, and must keep them always erect and spread, even at rest. Their flight muscles are attached at each side of the base of each wing, giving the wings a relatively small arc and a wing-beat rate of only about thirty times per second. Yet extremely strong flight muscles and the position of the wings near the center of balance make dragonflies capable of amazing acrobatics. They can shoot straight up, hover, dart sideways, accelerate to as much as thirty-five miles per hour, change directions in midflight, stop instantly, and fly backwards.

Dragonflies have excellent eyesight, due to large, protruding eyes with as many as 28,000 individual facets facing in all direc-tions, which allow them to spot mosquitoes, gnats, flies, and other small flying insects from as far away as forty yards. Their forward-thrust legs are located in a cluster near the front of the thorax, and are arranged in a basketlike way to catch prey and quickly transfer it to the mouth. All in all, dragonflies are so well adapted as air-borne predators that their legs are nearly useless for walking.

Because of their fierce appearance, dragonflies, though harmless, have been granted all sorts of nasty abilities. Wishful thinkers once thought they were capable of sewing shut the mouths of men who cursed and women who scolded, or of sewing up the ears of peo-ple who enjoyed gossip. According to a legend on England's Isle of Wight, dragonflies guide boys who have been good to places where fish can be caught, while boys who have been bad are stung. The Japanese consider the insects lucky and a symbol of victory in battle, and are always pleased to see them flitting above gardens. On the coast of the South China Sea, however, they are considered the unlucky omen of a coming typhoon.

A connection with violent weather may have some basis in fact, at least in the case of a South American species that sometimes migrates in large numbers before the powerful and cold pamperos winds that periodically sweep the plains of Argentina. According to an account by the naturalist W. H. Hudson, those dragonflies "make their appearance from five to fifteen minutes before the wind strikes; and when they are in great numbers the air to a height of ten or twelve feet . . . is all at once seen to be full of them, rushing past with extraordinary velocity in a northeasterly direction. . . . Of the countless millions flying like thistledown before the great pampero not one solitary traveler ever returns." Other migrant dragonflies include an Australian species that breeds in temporary pools then migrates in search of other pools, and is so willing to travel long distances to find them that it has been seen in a swarm over the ocean 900 miles off the coast of Australia. The green darner of North America migrates north into Canada each summer, where it breeds. The offspring fly south when cold weather arrives in September and October, and can sometimes be seen riding thermals and tailwinds along with migrating hawks and songbirds.

The hum of a mosquito somewhere in the room is so noticeable in the middle of the night because a mosquito is a buzzing dynamo: It beats its wings 600 times each second. It and the other members of the Diptera, or two-winged flies, are the speediest, most vigorous, and agile flyers of all insects. Horseflies routinely reach thirty miles per hour, and can probably exceed forty. Other flies can hover for long periods, and fly backwards. Among the world's most impressive stunt-fliers is one parasitic fly with the ability to lay eggs in the abdomen of an airborne bee.

Flies fly so well because of a curious adaptation of their wings. Instead of the double wings of many insects, they are equipped with only one pair of working wings. The rear set evolved into small, specially modified structures known as halteres. Halteres are shaped like tiny clubs or long spoons attached by their handles to the thorax behind the front wings. In flight the halteres oscillate rapidly, like wings, but instead of propelling the insect, they transmit subtle information to the fly to help regulate its flight, telling it to adjust for wind and how to perform the complex motions necessary to land upside down, say, on a ceiling.

If ever a thing with wings inspired poets to lofty sentiments, it is the butterfly. To Walt Whitman, in *Specimen Days*, they are dipping and oscillating, giving a curious animation to the scene. The beautiful, spiritual insects! straw-color'd Psyches! Occasionally one of them leaves his mates, and mounts, perhaps spirally, perhaps in a straight line in the air, fluttering up, up, till literally out of sight. In the lane as I came along just now I noticed one spot, ten feet square or so, where more than a hundred had collected, holding a revel, a gyration-dance, or butterfly goodtime, winding and circling, down and across, but always keeping within the limits.

The delicate, fragile beauty of butterflies in flight inspired the ancient Greeks to represent Psyche, the human soul, as a butterfly (and in turn inspired entomologists to give the name *Psychidae* to a family of moths). A legend of the Pima Indians of North America told of the Creator flying around the earth as a butterfly in search of a place for humans to live. In northern Asia a butterfly or moth was tied around the neck of an ill person, to replace the winged soul that escaped the body and caused the illness. The Irish believed that a butterfly flitting above a corpse indicated eternal bliss for the deceased soul. In Christian theology a man or woman must pass through two stages—life, symbolized by a caterpillar, and death, symbolized by a chrysalis, or cocoon—before he or she can be resurrected as a fully winged, gloriously painted butterfly. In China the butterfly was long symbolic of longevity and was a favorite subject of artists and poets. Less happily, butterflies were considered the souls of witches in parts of Scotland and have been considered evil omens in Cambridgeshire, Bulgaria, and in the Russian Ukraine. Wherever it is found in Europe, the death's head moth, with skull-like markings on its thorax and the unusual ability to emit a chirping sound, has been considered sinister.

Butterflies and moths are classified as the order Lepidoptera, from the Greek for "scaled wing." The scales, which rub off as fine dust on your fingers, are arranged in overlapping patterns, like shingles on a roof, and serve to protect and color the wings. They are almost unbelievably elaborate, and so tiny that each square inch of the Brazilian morpho's wing, for instance, contains 165

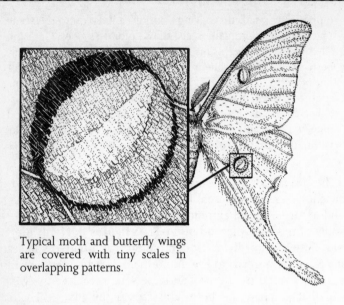

Typical moth and butterfly wings
are covered with tiny scales in
overlapping patterns.

rows of scales, with 600 scales in each row, for a total of 99,000
scales.

Few creatures are as adapted so completely for flight as butterflies and moths. Butterflies, especially, have disproportionately
large wings—a butterfly with a body weight about the same as a
bee's has wings twenty times larger—and generally have the highest ratio of wing-surface to body weight of all insects. They appear
to be *all* wings, with a narrow hinge of body joining them together.
Their flight is slow, erratic, and nearly effortless, in contrast to the
speedy flights of some moths, such as the hawkmoths, which have
long and narrow wings that make them extremely fast.

Insect wings are among the engineering marvels of the world.
They have evolved into complex structures of incredible strength,
flexibility, and subtlety. Insects rarely glide, the way most birds do,
so they must flap their wings almost continuously to generate lift
and stay airborne. In flight, they can contort and adjust their wings
in a variety of ways to make them more efficient. They twist them
to vary the "angle of attack" during each stroke, creating an aerodynamic effect similar to a curved propeller blade. They adjust the

wing's camber—or convex curve from leading edge to trailing edge—to increase or decrease the amount of lift. They alter the amount of surface area exposed to the air, by, for example, increasing the overlap between the forewings and hindwings. They generate more force by accelerating the speed of the downstroke while slowing the speed of the upstroke. And they can do all those things simultaneously.

No wonder dragonflies escape swallows and houseflies evade swatters. In the world of flight and fliers, insects are often looked down on as lesser, simpler creatures than birds and bats and astronauts. Those of us who want to soar with the eagles might do well to think smaller.

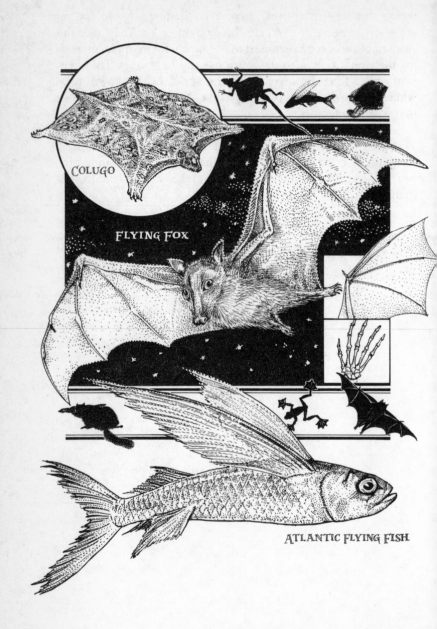

COLUGO

FLYING FOX

ATLANTIC FLYING FISH

Mammals, Fish, and Other Animals in Flight

The sky is not the domain only of birds, insects, and airplanes. Every class of vertebrates has members that fly or at least glide through the air to avoid predators, chase prey, or travel more efficiently from place to place.

The only animals other than insects and birds capable of true flight are bats. The 1,000 or so species worldwide make up a large order known as *Chiroptera*, or "hand-wing," in reference to the fact that the bones in a bat's wings are essentially the same as the bones in human hands. Bats range from the size of bumblebees to giant fruit bats and flying foxes with wingspans up to six feet. About 70 percent of all bat species are insect eaters; most of the rest dine on fruit or nectar. A few eat frogs and lizards, at least one eats fish it scoops from lakes and rivers with its clawed feet, and one South American species feeds on blood it laps from sleeping mammals.

A bat's wings are composed of a thin membrane covering a framework of bones anatomically similar to human hands, which gives the bat more agility in flight than even swifts or hummingbirds. They are not the fastest of flyers, but they are so nimble that at top speed they can turn 90 degrees in a distance less than their own body length. The large wing surface creates immense lift, enabling bats to carry their nursing young and up to twice their body weight while in flight.

The most celebrated skill among the insect-eating bats is echolocation. While hunting, a bat sends out up to 200 high-pitched squeaks per second through its nose or mouth, depending on species. When the sound waves intercept an object, whether a tree trunk or a flying mosquito, they bound back from it to the bat, which then processes the information. If the object ahead is interpreted as an insect, the bat swoops ahead and snatches it in its mouth or scoops it into a membrane stretched between its hind legs, cupping it there in the pouchlike opening and reaching in to eat it as it flies. Scientists studying echolocation once estimated that a bat's sonar was about a billion times more sensitive than the best sonar or radar yet devised by humans. The system certainly works: During an ordinary night of feeding a bat can eat as many as 3,000 insects. A single large colony of free-tailed bats in Texas is estimated to capture and devour 250,000 pounds of insects every night during the summer.

The high-pitched tone of the sonic squeak is inaudible to most other animals, but a few species of underwing moth have developed ears sensitive enough to hear it. When an approaching bat turns on its sonar some of those moths immediately begin frantic evasive action. Others take their defenses a step further by emitting high-pitched tones of their own that serve to "jam" the bat's sonar.

Few animals have been burdened by as many misconceptions as bats. They do not, for instance, tangle in women's hair. They are no more likely than other small mammals to be carriers of rabies and are even less likely to be transmitters of the disease since the teeth of most bats are not strong enough to puncture skin. Bats do not live only in caves, and, far from being blind, the eyesight of many species is excellent.

Caves are important sanctuaries, however, to many species of hibernating bats. Beginning in the fall, when temperatures drop and the insect population begins to decline, they are capable of the quickest and easiest hibernation in the animal world, their hearts slowing in a very short time from 180 beats per second to three beats per minute, and their respiration slowing from eight breaths every second to eight every minute. With their bodies stocked with accumulated fat, and gathered together in colonies of hundreds or thousands of individuals to utilize their combined body heat, bats are capable of living for months in hibernation.

Bats are the only true flying mammals, but some other mammals have developed the next-best thing. The "flight" of a flying squirrel might not be true flight, but anyone who has watched the perky little mammal's swooping descent from a treetop to a bird feeder will give it points for trying. The difference between flight and gliding is mostly a matter of degree: A glider does not get vertical lift, but rather delays its fall by making itself as broad and flat as possible to take advantage of the density of air.

The loose fold of furred skin between the front and hind legs of flying squirrels and other gliding mammals is called a *patagium*. A flying squirrel glides by carefully measuring the distance to a target, then leaping from a tree trunk or other height, spreading the patagium taut, and gliding down and forward with its body horizontal and tail straight out behind, tilting slightly to adjust direction. Just before reaching its destination—usually another tree trunk—it swings its body up to a vertical position, climbs slightly in the air, and lands with all four legs simultaneously. Typical gliding distances are 130 to 165 feet, but some individuals have been seen to travel as far as 660 feet. Different species of flying squirrels are found in North America, Europe, India, China, and Indonesia. The largest, with a thirteen- to nineteen-inch body and eleven- to fifteen-inch tail, is the white-cheeked giant flying squirrel of Japan and western China. Others include the lesser American flying squirrel, the Eurasian lesser flying squirrel, and the brown giant flying squirrel. All are primarily nocturnal and feed on nuts, seeds, and occasional insects and young birds.

Australia and New Guinea are home to several species of marsupial gliders. The greater and lesser gliding possums, members of the genus *Petauroides*, are similar to flying squirrels, with their furry gliding membranes and long bushy tails. They are nocturnal woods-dwellers that feed on nectar, sap, insects, and small vertebrates, and have been seen to glide up to 165 feet between isolated eucalyptus trees.

The most skillful of the mammal gliders are probably the rabbit-sized colugo of Indochina and the Philippine gliding lemur, creatures so unique the two species belong to an order all their own. Although their gliding membranes are similar to those of flying squirrels, they are much larger, considerably larger than the animals themselves, so that when they are walking the folds of skin look like furry cloaks thrown over their backs. While gliding, the patagium spreads out more like a parachute than wings, and allows the colugo and gliding lemur to glide more than 400 feet. One individual was observed gliding between tree trunks: In a slow glide of 445 feet it lost only 35 feet of elevation.

Flying fish, members of the family *Exocoetidae*, have broad pectoral fins they can extend out from their bodies like wings. When startled or pursued by predators these inhabitants of tropical and subtropical seas swim at speeds up to about forty-five miles per hour, their tails beating 80 to 100 times per second, then leap from the water, spread their wide pectoral and ventral fins, and glide hundreds of feet. For extended journeys, they lower their tails, beat furiously, and continue across the surface of the water. Most glide four or five feet above the water, but they on occasion rise high enough to land on the decks of ocean-going ships.

The Indonesian gliding lizard, *Draco volans*, or flying dragon, has a pair of gliding membranes extending from its ribs, which it keeps folded when not in use. A rainforest dweller, it is often seen clinging to a tree trunk and looking no different than any other small lizard (it measures up to about nine inches, but most of that length is tail). When alarmed, however, it launches into an evasive glide, its brightly colored and spotted "wings" spread out like butterfly wings. It glides fifty to sixty-five feet under most circumstances, but is capable of distances up to 330 feet.

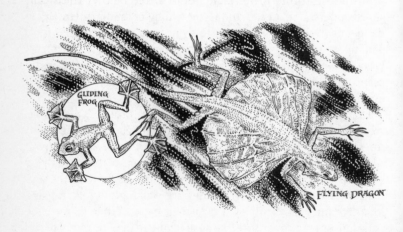

In Malaya and Burma are two species of tree frogs, Wallace's gliding frog and the Malayan gliding frog, capable of gliding forty or fifty feet between trees. Like all tree frogs, both species have sucker discs at the bottoms of their feet, but they are much broader than usual and the toes are fully webbed and much longer than those of other frogs. When the frog leaps, each foot's spread webbing acts like a small parachute. The body, with its four legs outspread and underbelly drawn up and held in a concave shape, increases the gliding surface.

Psychologists estimate more than half of all people feel at least some fear of snakes. If so, then the paradise tree snake of Malaya and the golden tree snake of southeast Asia are the stuff of nightmares. These three- to five-foot long dwellers of the forest canopy have evolved the ability to draw in their bellies, extend their side scales, and, with bodies flattened, glide from the tops of trees to the ground. Western zoologists who first heard about them were naturally dubious, especially since the stories included local folklore and legends that described the snakes changing into birds when they wanted to fly, then changing back to snakes when they landed (and eating the bird). Finally an openminded naturalist named Major Stanley Flower captured a specimen, and after patient prodding succeeded in making it glide from the top window of his house to the lower branches of a tree below. The world rejoiced.

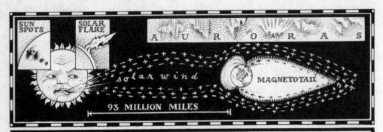

AURORA NIGHTS

On a crisp autumn night in 1967, the sky above our house danced and shimmered with mysterious lights. My father woke my brother and me, urging us to dress quickly and hurry outside to the yard. We stood in the cold, wrapped in blankets, and saw the sky to the north flare with enormous shafts of light, like the grand opening of the greatest shopping center on earth. As we watched, the lights rippled into slowly waving curtains, the color changing from white to the faintest pink, then became viscous, each shaft twisting and undulating in rhythm like broad underwater weeds in current. Suddenly the shafts of light began to climb, higher and brighter, until they rose to the very top of the sky and became the most awesome sight of my young life.

As a child I understood the Northern Lights to be caused by sunlight reflecting off ice at the North Pole, an explanation my friends and I found strangely satisfying. Not surprisingly, these awesome natural phenomena have inspired even more imaginative explanations. Some Eskimos and Tlingit Indians, for instance, believed the lights were caused by the spirits of the dead playing a game similar to soccer, using the skull of a walrus for a ball. Norsemen long ago thought they were seeing the reflection of sunlight off the shields of the Valkyries—those warrior maidens who transported slain heroes to Valhalla. The Mandan Indians of the American plains theorized they were the glow from the fires of northern tribes boiling their enemies in enormous pots. In England, auroral displays, especially those colored red, were believed to prophesy war, pestilence, and famine. A happier interpretation appears in an Estonian folktale describing the aurora as an enormous wedding in the sky, attended by guests arriving in luminous sleighs and horses.

In regions where they are uncommon, auroras have been misinterpreted as everything from fires to UFOs. According to the Roman philosopher Seneca, emperor Tiberius Caesar witnessed a rare aurora early in the first century A.D., assumed the colony of Ostia was in flames, and hurriedly dispatched a column of soldiers to their aid. A particularly brilliant display in England in 1938 caused the fire department to be mobilized in the belief that Windsor Castle was burning. In 1941, residents of Washington, D.C., decided an aurora was either a secret weapon being field-tested by the army or, worse, searchlights warning of an impending German air attack.

Recent scientific theories are considerably less colorful than the legends and fallacies, attributing both the aurora borealis of the Northern Hemisphere and the aurora australis of the Southern Hemisphere to the interaction of charged particles in the atmosphere.

Observer's have long noted that aurora activity coincides with disturbances on the surface of the sun. Sunspots and solar storm activity tend to peak about every eleven years (although it can vary from as much as 7.5 to 16 years), creating explosions of gases called solar flares so intense they break up atoms of gases into negatively charged electrons and positively charged protons. Propelled by the solar blasts, enormous amounts of those magnetic particles shoot out in all directions into space, creating a "solar wind" fast enough to reach the earth in about two days. As the particles stream past the earth they behave much the way river current does as it flows around a midstream rock, circling behind it in an eddy. In the earth's case the eddy is known as the magnetotail, and the current is drawn back at tremendous velocity to the magnetic fields of the north and south poles. It is there, 60 to 200 miles above the earth's surface, pulled into the atmosphere by magnetic attraction, that the charged particles from the sun collide with atoms and molecules of air to generate the vast electric glow we see as the aurora.

Solar flares distort the magnetic field of the earth and cause the auroras to move farther from the poles. That is why during years of exceptional sunspot activity the Northern Lights are likely to be visible much farther south than normal—as far south as Mexico,

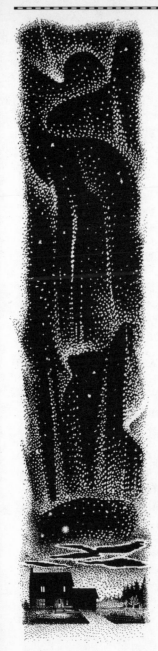

on occasion. More frequent sightings occur north of a line from New York City through Pittsburgh, Des Moines, Boise, and Salem, Oregon. Curiously, the best place on earth to view the Northern Lights is not, as you might think, the North Pole. The phenomenon is visible most often—about 243 nights each year—in an oval-shaped band running through northern Norway, central Hudson Bay, Point Barrow, Alaska, and northern Siberia.

On the best nights auroral lights occur as lazily shifting bands, rays, crowns, arcs, or draperies, and can appear in brilliant shades of red, yellow, green, blue, and violet. Colors are caused by the solar particles reacting with various atoms and molecules. Oxygen and hydrogen, for example, glow red or green, and nitrogen glows pink or purple.

The Northern Lights are most common during September and March, when cool, clear nights enhance visibility. Look for faint vertical bars of light low in the northern sky. Dim displays are easily mistaken for artificial lights, like the spotlights that illuminate billboards, but an aurora, even a faint one, will shift, dim, brighten, and undulate if watched for a few minutes or more. A faint aurora will sometimes flare brightly for a few moments, then fade altogether, or it might expand into a sky-filling display. Expect the best displays around midnight.

And be sure to wake the kids.

Does the Aurora Roar?

FOR YEARS EYEWITNESSES have claimed that in the polar regions of the world auroras can sometimes be heard making crackling or swishing noises, a kind of cosmic static. In 1933, Science magazine published this account by Clark M. Gardner, who claimed to have heard an aurora:

> In my own mind there can be no doubt left as to the audibility of certain types of aurora, for I have heard them under conditions when no other sound could have been interpreted as such, for no other sounds were present . . .
> In the winter of 1925–26 I was engaged in making a drive of reindeer across the mountain range bordering the Arctic coast north of Cape Prince of Wales on Bering Strait. . . . As we climbed with our dog team at the summit of the divide we were both spellbound and astounded by the magnificent display of aurora, the most wonderful display I have ever witnessed during my eight years of life among the Eskimos. . . .
> As we sat upon the sled and the great beams passed directly over our heads they emitted a distinctly audible sound which resembled the crackling of steam escaping from a small jet. Possibly the sound would bear a closer resemblance to the cracking sound produced by spraying fine jets of water on a very hot surface of metal. Each streamer or beam of light passed overhead with a rather accurate uniformity of duration. By count it was estimated to require six to eight seconds for a projected beam to pass, while the continuous beam would often emit the sound for a minute or more. This particular display was so brilliant that traces could easily be seen long after daylight.

It was theorized that the aurora's sweeping fields of static electricity might be audible in a remote environment where silence was nearly total. In addition, Eskimos were known to refer to certain auroras as "noisy," seeming to add credibility to the tale.

Skeptics are doubtful. An American scientist studying the ionosphere during World War II learned the Eskimo language well enough to ask about the references to audible auroras, and was told by the Eskimos that they used the word "noisy" only as a figure of speech. If an aurora display was especially bright they referred to it as *neepeealo*, or "very noisy," reasoning that such brilliant banners would have to make noise far up in the sky, since they looked like clothes flapping in the wind.

Most scientists now believe reports of crackling and swishing noises actually result from the observer's own breath freezing in the still, frigid air of the arctic or antarctic night. Even if the charged particles of the auroras did make noise as they interacted, they say, it is extremely unlikely the sound would be heard 60 to 200 miles below on the surface of the earth.

PERIWINKLE

HORSESHOE CRAB

CALIFORNIA GRUNION

MOONSTRUCK

There is something haunting in the light of the moon; it has all the dispassionateness of a disembodied soul, and something of its inconceivable mystery.

—Joseph Conrad, *Lord Jim*

The moon is full tonight, swollen and bright and bleating with strangeness. I know it is full because the calendar says so, but the moon itself is hidden behind a dense mantle of clouds. If I was receptive enough perhaps I would know the phase of the moon in my blood. I might feel gravity pulling at the saltwater in my veins the way it pulls at the saltwater in the Bay of Fundy. I might grow unaccountably restless, pace the floor, howl, dance, fall in love.

Living indoors as most of us do, muffled by so many generations of reason and civilization, we seem to have grown immune to most of the moon's magic. Were we less rational we might suffer bouts of lunacy brought on by sleeping in the open beneath a full moon—moon-struck in the original sense of the word—or find validity in tales of lycanthropy and bewitchment. The moon has been blamed for everything from bad moods to traffic accidents to earthquakes, but whenever someone proposes a theory to explain how that might be so, someone else steps forward to demonstrate why it does not stand up to skeptical inquiry.

Still, romantics and skeptics alike agree that the moon has always been a potent force on earth. So pervasive are lunar influences that the moon has come to be associated with the mysteries of love, sex, magic, birth, and creation in virtually every known culture. Many mythologies have characterized the moon as a goddess, and have bestowed her with mysterious, arcane powers. Diana was the ancient Roman goddess of the moon and the protectress of women. As sister of Apollo, god of the sun, she was associated with forests, animals, hunting, and childbirth. The Greek goddess Artemis, virgin daughter of Zeus and Leto, was worshipped as a moon goddess with strong ties to wildlife and hunters. The moon goddess Selene was sister to the sun god Helios, and was a significant contributor to legends of sorcery. Hecate, another Greek goddess of the moon, was the patroness of sorcerers and witches and the ruler of storms.

Because of such strong associations with witchcraft and sorcery, it is not surprising that lunar deviations have often been perceived as ill omens. In medieval Europe, lunar eclipses inspired great fear. Even the relatively common occurrence of the old moon being faintly visible inside the cusps of the new moon was reason to believe that people would come to harm, as recounted in "The Ballad of Sir Patrick Spens." The phenomenon described in the ballad—"I saw the new moon / Wi' the auld moon in her arm"— is caused by earthshine: sunlight reflecting off the earth which in turn reflects off the unlit surface of the moon.

Several theories have been proposed to explain how the moon originated. One is the so-called daughter theory, that the moon was formed when the earth, rotating much faster than it does now, became elongated and a portion of it was torn from what is now the Pacific Ocean. The spouse theory proposes that the moon was a passing asteroid or other body attracted by the gravitational pull of the earth. Finally, the sister theory argues that the earth and moon were formed at the same time, when clouds of cosmic matter first condensed into the planets and moons of the solar system.

Whatever its origin, the moon exerts a powerful and inescapable pull on our planet. It does not, properly speaking, rotate around the earth. The two, like dancers clasping hands and twirling, rotate around a center of gravity known as the barycenter, which is located beneath the earth's surface, about 3,000 miles from its center. The moon, with only $\frac{1}{81}$ of the earth's mass, pulls so inexorably on the planet that the oceans bulge, creating the tides. Pliny believed "lunar energy penetrates all things." Aristotle noticed that the ovaries of sea urchins swell during the full moon, and Cicero believed that populations of shellfish increased and decreased according to the phases of the moon. More recently it was discovered that shore organisms like flatworms and periwinkles, when placed under laboratory conditions, continued rhythmic activity that coincided with the cycles of the tides. Oysters taken from Long Island Sound in 1954 and shipped to a lab 1,000 miles away in Evanston, Illinois, maintained a cycle of valve openings and closings that at first corresponded to the tidal rhythms of their home waters; gradually, however, those rhythms adjusted to coincide with the lunar schedule in Evanston.

Of the many marine animals that have synchronized breeding activity based on lunar-tidal cycles the most famous is probably the California grunion. These six- to eight-inch fish time their breeding cycle to come ashore on waves during the first three nights after the peak of the spring—or highest—tides. Once stranded by a receding wave the females shimmy tail-first into the wet sand to discharge their eggs, and the males soak the sand above with their milt. The adults return to sea on the next waves and the eggs are left to develop in the sand, then hatch when waves cover them again two weeks later during the next high tide.

Lunar influences on birds and mammals are less obvious, and certainly more controversial. A 1930 study of European nightjars found that eggs were laid during the last quarter of the lunar cycle, so that the chicks would hatch by the next full moon when the adults could hunt all night to feed them. Darwin, in The Descent of Man, observed that "Man is subject, like other mammals, birds, and even insects, to that mysterious law, which causes certain normal processes, such as gestation, as well as the maturation and duration of various diseases, to follow lunar periods." A 1959 study by Walter and Abraham Menaker, relying on a massive data base, found the average human menstrual cycle to be 29½ days—precisely the length of the lunar month. The same study, based on 250,000 births, found the average length of human gestation to be 265.8 days—nine lunar months—leading the Menakers to conclude that the human reproductive system follows lunar time, not solar time. Most researchers today remain cautious about accepting that conclusion. W. B. Cutler, in the *American Journal of Obstetrics and Gynecology*, went further than most, admitting to the "possibility of a relationship" between lunar phases and menstrual cycles, with the moon perhaps affecting electromagnetic fields that in turn might influence menstruation in some women.

Brazilian mothers hide their newborn children from the moon, and in Iceland a pregnant woman will not sit facing the moon for fear her child will be born a lunatic. The notion that the moon influences mental health is ancient, going back long before *Othello*, where Shakespeare wrote:

> It is the very error of the moon;
> She comes more near the earth than
> she was wont,
> And makes men mad.

Charles Hyde, the English laborer who served as the model for Robert Louis Stevenson's Jekyll and Hyde character, committed his criminal acts primarily during the full and new moons and argued in court he was not responsible for his actions because of "lunacy." (The court was not impressed: He was sentenced to prison in 1854.)

Frequent claims by police officers and hospital workers that the incidence of violent crimes and accidents increases at the full and new moons have never been adequately substantiated. Studies of homicide records over several decades in various U.S. cities convinced Arnold Lieber and a few other psychologists in the 1970s that there was a statistical basis for such beliefs, and led Lieber to hypothesize that "biological tides" influence movements of salt, water, and hormones in the brain, resulting in all manner of bizarre behavior. Later studies, however, using larger data bases, failed to find any statistical evidence that the number of violent crimes varied with lunar cycles. Likewise, attempts to link the moon to incidences of schizophrenia, depression, suicide, arson, rape, automobile accidents, and epileptic seizures have been discarded as inconclusive and unscientific.

In spite of the controversy and the lack of definitive data, it seems likely that all life on earth is bound at least to some degree with the moon. According to one current theory, the moon at early stages in its history followed an orbit much closer to the earth than the present one. If true, the earth's oceans would have experienced tides far higher and more powerful than those we know, sweeping the shores of the seas like a beater in a vat, stirring chemicals into the combinations that led to the first forms of life. Organisms evolving under those tidal influences would have been subjected for hundreds of millions of years to the ebb and flow of massive tides. If we and most other forms of life on earth evolved from those creatures, it is conceivable, as Darwin believed, that some of our biological processes would still be affected by the moon-induced rhythms of the sea.

Modern wolf researchers have pretty much killed the myth of wolves howling lustily at the moon, but the fact remains that predators and prey alike are made restless and edgy by the pale light of the moon. Not so long ago—in the global scale of time—men, women, and children were prey as well as predators and would have learned that nights of full moon were a time to be alert and on guard. Living among tigers with 10-inch canines and cave-dwelling bears that would have made Kodiaks look like koalas, the moon must have represented far more than a mysterious object traced with odd, facelike images. The lessons it taught would be terrifying and unforgettable.

Innate or learned, that old devil moon still moves us. Even in these post-*Apollo* decades we remain awed by the sight of a rising moon. That we can still be mystified, with footprints etched as permanently as petroglyphs on its surface, suggests the moon's power goes deeper than the sentiments of popular songs and flowery poems.

A person no doubt has to be sensitive as a seismograph to register the full range of lunar influences, and I'm afraid I'm constructed of too coarse a material. One night last summer, with the full moon near its zenith and the woods settled into the rich, buzzing alertness of deep night, I walked a ridge of maples above the river near my home. I was aware of no primitive urgings, did not feel the obstinate swell of a tide in my veins, was in fact thinking of home and a cold beverage and bed when something com-pact and terribly fast exploded from a dark clump of underbrush at my feet, then crashed away through the bushes with all the delicacy of a flung bowling ball. My primal memory, I suspect, has lapsed—I saw no images of rushing wings and clenching talons, no claws or saber teeth—but it was interesting to observe how well my fear reflex works. Aided by moonglow or not, that rabbit scared the living daylights out of me.

Viewing the Moon

AUTUMN IS NOT the only time of year when the moon can be observed, of course, but there are good reasons to associate it with this particular season. Harvest moon, hunter's moon, the ghostly orb of Halloween—the moons of autumn are among the most memorable of the year. That is due partly to ancient rituals passed down from ages when lunar cycles were followed closely in the planting and harvesting of crops. The harvest moon is the full moon that occurs nearest the autumnal equinox and is named for the tendency for the moon at this time of year to continue rising early in the evening for several nights before and after full phase. Farmers for centuries have taken advantage of the extra illumina-tion and worked late to harvest their crops. The same thing hap-pens a month later, in October, when the full moon is traditionally called the hunter's moon.

There are practical reasons as well for thinking of the moon in autumn. Cooling air and changing wind patterns have swept away the humid, sullied atmosphere of summer. The moon we see on a cold October night is in sharper focus, more detailed, its cusps—if it is at half moon or less—sharp as scimitars. It is possible to see craters and mountain ranges with only a pair of binoculars, while startling details of the moon's surface can be brought into focus with a relatively inexpensive reflecting telescope.

For an event that occurs with regularity every month of our lives, the lunar cycle creates a lot of confusion. It helps to remember that the moon travels much slower than it appears when we watch its progress across the sky on a moonlit night. If you have ever tried photographing a full moon you know how quickly it moves. By the time the aperture and shutter speed are set and the lens focused, the moon is edging out of the frame. That apparent speed, however, is caused not so much by the movement of the moon as

PHASES OF THE MOON

1 NEW MOON
2 WAXING CRESCENT
3 FIRST QUARTER
4 GIBBOUS MOON
5 FULL MOON
6 GIBBOUS MOON
7 LAST QUARTER
8 WANING CRESCENT

Sunlight

by the movement of the earth. If the moon spun completely around the earth once every twenty-four hours or so, as it appears to do, then we could watch it go through several phases all in one night. In truth, however, its relative position to the earth and sun changes very little in one night. It is we who have changed, spinning rapidly on our axis, and everyone on earth sees the same phase of the moon.

The best way to appreciate the moon is to watch its progress through an entire lunar cycle, or synodic month, of 29 days, 12 hours, and 44 minutes. In that time it goes from new moon (when it is between the earth and the sun and virtually invisible because it reflects no sunlight), to full moon (when it lines up on the opposite side of the earth as the sun and reflects sunlight off its entire visible face), and back to new moon. It waxes as the illuminated portion grows larger after new moon, then wanes as the illuminated portion decreases in size after full moon. In the process it goes through crescent phases (less than half illuminated), quarter moon (half illuminated), and gibbous phases (more than half illuminated, from the Latin word for "a hump" or "bulging"). The half moon occurring after the new moon is called the first quarter, and the half moon occurring after the full moon is called the last quarter.

The new moon rises with the sun and sets with the sun and is visible only during an eclipse. A few days after new moon, a thin crescent rises shortly after sunrise and is visible as it follows the sun toward the west. It becomes brightest after the sun sets, and the sky begins to darken.

Each day the moon waxes larger and rises an average of 50½ minutes later, so that by first quarter, when it appears to be a half moon, it is straight overhead as the sun sets, and disappears over the western horizon at about midnight. From half moon to full moon it passes through gibbous phases, bulging and growing into a more and more nearly circular shape.

The full moon rises in the east as the sun sets and travels all night across the sky, setting in the west as the morning sun rises. After full moon, the waning moon continues to rise an average of 50½ minutes later each night. By last quarter the half moon is straight overhead as the sun rises in the morning, and is setting in the west at about noon. It shrinks each successive night into a smaller crescent until it disappears entirely at the new moon.

Because the synodic month of 29 days, 12 hours, and 44 minutes is shorter than most of our calendar months it happens that every 2.7 years we can have two full moons in the same calendar month. The second one has traditionally been known as a "blue moon." Thus, "once in a blue moon" is an event that happens rarely. A literal blue moon can be seen occasionally in winter, when sunlight and moonlight are dispersed in the upper atmosphere, illuminating the blue end of the color spectrum.

We always see the same side of the moon. The hidden or dark side was the subject of much speculation before the first space probes returned with photographs showing it differed from the lit, familiar side only in being marked with more craters, a result of the increased number of meteorites striking it on that side, and of having fewer seas. The moon revolves on its axis once each month, always keeping the same face toward the earth because of a slight bulge that has been captured by earth's gravity. We can see a little more than half the moon, however—about ⅗ths of it—due to librations: variations in the moon's motion that cause it to swing slightly one way then the other.

According to Aristotle's concept of the universe, considered gospel until the seventeenth century, the moon and all other heavenly bodies were infinite, perfect, and incorruptible—only earthbound things were changeable. Thus the moon had to be perfectly smooth, without irregularities. Galileo challenged that view when he looked at the moon with a telescope for the first time and declared it had mountains and seas. A similar view is available to anyone with a simple telescope or even a pair of binoculars. The mountains are clearly visible, and what Galileo understandably interpreted as oceans we now recognize as large, flat plains, though they retain their Latin designation as maria, or seas.

Watch the moon, night after night, month after month, and it becomes evident why it is one of the most celebrated sights in literature and music. It is celebrated not only because it is beautiful, but because it is strange and mysterious and difficult to comprehend. It has often been associated with the intuitive, instinctive, and nonverbal faculties of the mind, at polar extremes to the logical, lineal faculties associated with the sun. That we arrange our lives according to a complicated and not-quite-precise calendar based on both lunar and solar cycles suggests how important both ways of thinking are to us.

Why Does the Moon Appear Larger Near the Horizon?

THAT ENORMOUS FULL MOON rising above the horizon has puzzled moon watchers for centuries. As it climbs it seems to shrink—as much as 25 percent, in most people's estimations—in spite of the fact that when it is directly overhead the moon is actually closer to an observer than when it is viewed low on the horizon.

The distortion of the moon's shape (like the sun's) when it is low on the horizon can be blamed on refraction through the earth's atmosphere, but it does not explain why both the sun and moon appear larger. One theory proposed long ago to explain the phenomenon is that the moon only appears larger because we unconsciously compare it to nearby objects like a row of trees or a mountain on the horizon. While that makes logical sense it does not explain why the moon and sun appear large even when rising over the sea. Another theory suggested that because we perceive the sky as a domed vault, we unconsciously think of the horizon as farther away than the "top" and make the moon larger on the horizon to compensate for the difference. Still other researchers have discovered that for unclear reasons the mind tends to make things seem larger when viewed horizontally than when seen vertically.

Some nights the moon appears slightly brighter because its orbit is elliptical, causing it to be sometimes closer, sometimes farther from earth. At its closest, or *perigee*, it is about 27,000 miles closer than at its farthest, or *apogee*. When the full moon coincides with the perigee of its orbit, tides are higher than normal, and people, perhaps, are slightly more prone to lunacy than usual. It might be a time, as Milton described in *Paradise Lost*, of "maladies of ghastly spasm . . . moping melancholy and moon-struck madness."

MOTEL
VACANCY

TYPES OF FOG

RADIATION.   ADVECTION

COOLING LAND

WARM AIR
cool water   land

UPSLOPE   SEA SMOKE

WARM AIR
mountain   water–5°–10° warmer

cool air

OBSCURED BY DEGREES: FOG, SMOG, HAZE, AND DEW

The fog comes
on little cat feet.
—Carl Sandburg

I have never seen fog that looks even remotely like pea soup, but one autumn night driving along a coastal highway in Oregon, my wife and I found ourselves in a soupbowl of fog so thick we could not even see the road in front of us. The road was winding and hilly and bordered by precipices of unknown height, but to stop in the middle of that dense white nothingness was to risk being hit from behind by less cautious motorists. The only thing to do was drive on, Gail with her head out the passenger window watching the tiny cone of highway beneath the headlight and telling me when to nudge left or right to keep the wheel on the white line. We traveled perhaps three miles in one anxious hour before pulling over in the murky glow of a motel's neon sign. The fog had indeed seemed as thick as soup, the kind you can stand a spoon in.

Carl Sandburg was far more appreciative of fog than the average fog-bound motorist or airline passenger. Children are not especially impressed with the small world that circles them in a fog and are invariably disappointed to learn that being inside a cloud is little different. Looking down on a valley brimming with a dense river of fog is far more pleasing than being in the fog itself, just as watching clouds from a distance is more satisfying than watching the world grow obscured through the porthole window of a cloud-flying aircraft. Most of us still like to think that clouds are sturdy enough to support our weight. Fog reminds us how insubstantial they really are.

But fog is an excellent demonstration of the way water and air act when they are placed in an intimate relationship. The tiny droplets of water contained in fog are visible, or so nearly visible that they seem to hover on the border of visibility. Held together in a vaporous cloud, they illustrate condensation in progress, water in exquisite balance with air. The droplets suspended virtually weightless in a bank of fog are so eager to be born into water that they will coalesce at the slightest urging. Pass your bare arm through it and water clings to the hairs like dew glistening on blades of grass. Set a chilled glass in it and the glass becomes an instant condenser of water. Walk or run through it, and you can feel the droplets breathing coolly against your skin.

Fog is relatively common in every season in some places and can form anywhere or any time there is a combination of cool temperatures, moist air, and low or nonexistent wind. The most common type of fog, *radiation fog*, results when the ground cools and causes moist air above it to condense into water vapor. It is most common on the clear, calm, lengthening nights of autumn, when the ground loses heat and there is no cloud cover to trap the warmth as it escapes into the atmosphere. Such fog occurs most frequently in valleys, hollows, and other low-lying areas where cool air collects and just after dawn, when the first heat from the rising sun causes a mixing of warm, moist air with the cold air settled near ground. A slight wind helps form radiation fog by mixing more molecules of air and water vapor and putting them in contact with the cool ground. Because it is denser and heavier than the surrounding air, the fog tends to settle in valleys and other low areas. It often forms in the early evening, when winds die and the air cools, and lasts until rising temperatures and winds disperse it in the morning.

Advection fog is common along shorelines, and is essentially a ground-hugging stratus cloud created when a layer of warm, moist air condenses as it is carried over a cool surface of water or soil. Coastal regions are susceptible to fog because the air there tends to become heavily loaded with evaporated water. By late summer and early autumn the water is likely to be significantly warmer than nighttime air temperatures, causing water particles in the air to cool and condense into fog. Advection fog is generally more common on west coasts than east because prevailing winds carry moist air in from the water. It occurs frequently along the California and the North Atlantic coast, where warm air comes in contact with currents of extremely cold water. San Francisco's famous fogs are caused by humid, warm air drifting shoreward over the cold California current. On the opposite side of North America, the fogs of Canada's Bay of Fundy are particularly noteworthy, and are caused by the bay's shallow water, which is warmer than the North Atlantic that feeds it. Dense advection fogs are common in the North Atlantic and are caused by moist winds blowing north along the Gulf Stream and coming in contact with the cold Labrador currents in the waters around Newfoundland.

Upslope fog is a form of advection fog that occurs frequently along the sides of mountains, where currents of warm air follow upward-sloping land until they reach an altitude where cold surfaces cause condensation. New Hampshire's Mount Washington experiences upslope fogs about 300 days each year. They are common too in the Plains states where humid winds from the Gulf of Mexico climb the foothills of the Rockies and in coastal regions where moist sea air is forced to rise by hills.

Steam fog, also called "sea smoke" or "frost smoke," occurs over open water, especially in early autumn, when air settling to the surface is 5 to 10 degrees Fahrenheit colder than the water. The air immediately above the ocean, lake, or river picks up warmth and evaporated water, mixes with the cold air above it, and condenses into a thin wispy fog lingering a few inches to a few feet above the water. When it occurs during light winds, the fog trails from the tops of the waves.

Precipitation fog hovers low to the earth when warm raindrops fall through a layer of cold air near the ground. It occurs often during spring rains, when the remains of the winter snowpack keep a layer of refrigerated air above it.

Ice fog forms when steam or human breath comes in contact with air colder than −40 degrees Fahrenheit and condenses into ice crystals. It is rare except in the world's coldest places, such as the hot-water geysers in Yellowstone Park, and in Antarctica, where steam exhausted from camp stoves forms floating fogs of ice crystals dense enough, on occasion, to completely obscure a camp.

The fogs of London and San Francisco are the most famous, but there are other places where fog is even more common. The foggiest spot on earth may be Cape Race, at the southeast corner of Newfoundland; its view of the Atlantic is obscured by fog an average of 158 days each year. The record for the most fog on the West Coast of the United States is held by Cape Disappointment, Washington, where it is foggy nearly 30 percent of the time—an average of 2,552 hours of fog per year. Mistake Island, Maine, holds the East Coast record, with 1,580 hours of fog each year.

Hazy Days

UNLIKE FOG, which is usually a local phenomenon, haze affects large areas at once. It is most common on sultry summer days, when the air is too warm or contains too little water vapor to condense into clouds and there is no wind to blow away airborne particles. Haze is air that wants to turn into clouds, but can not.

Filmy, dusty, obscuring haze can blur the horizon and dull the sun because it is composed of an unusual abundance of gases and airborne particles, known collectively as aerosols. If conditions were right for condensation, many of those aerosols would serve as nuclei for droplets of water vapor. Instead, the particles linger in the atmosphere and scatter sunlight. The color of the haze is an indication of the type of particle predominant in that part of the sky. The grayish haze common to shoreline regions of the oceans is caused by minute particles of salt tossed into the air by waves and wind. The blue haze so frequently seen over the Blue Mountains of Australia, and the Blue Ridge and Great Smokey Mountains in the Appalachian Range of the eastern United States, is caused when ozone combines with hydrocarbons known as *terpenes*, which are given off by vegetation and rise on the upland slopes of wooded mountain ranges. In the Middle East, enormous quantities of fine dust blown up from the desert create a reddish haze distinctive enough to give the name the Red Sea to the body of water where the haze is most vividly reflected. The gray and brown haze suspended over metropolises and industrial centers is smog, a mixture of highway dust, automobile exhaust, oil fly ash, coal fly ash, and other human-made pollutants.

Atmospheric Inversions:
A Fuliginous and Filthy Vapor

THE MALEVOLENT GROUND FOG of low-budget horror films (which in 1979 reached a summit of sorts in John Carpenter's *The Fog*) is a cliché that keeps monsters conveniently out of sight, but it reveals nothing about the nature of fog. Is fog dangerous? Yes, if it obscures highways and airports, but no, not inherently. Still, there have been occasional claims of deadly fogs, such as the one reported in *Science* magazine in 1931, under the title "The Fatal Belgian Fog": "About the week-end of December 7, [1930] an extremely heavy fog prevailed in Belgium and England, and the daily press reported that in the neighborhood of Liege more than forty persons and a considerable number of cattle died, exhibiting symptoms of asphyxiation. Autopsies performed on twelve cows indicated that they had died from pulmonary edema. . . . Final judgment on this phenomenon must await the results of the investigation which the Belgian government has undertaken."

The investigators of that fatal fog discovered that it was not the fog itself that killed people and cattle, but industrial pollutants from mills and factories held down in a pool of stagnant air by an atmospheric condition called an *inversion*. The fog that had seemed at

DEW & SMOG IN QUEENS

first like a culprit was but a symptom of the unusual amounts of particulate matter in the air.

Human-made and natural pollutants are never good for the health of the planet and its inhabitants, but when caught beneath an atmospheric inversion they can be deadly. An inversion is created when a layer of warm air traps cooler surface air beneath it, preventing it—and the wastes it contains—from rising and dispersing. When it occurs in lightly inhabited regions, say over a lake surrounded by hills or mountains, there is little harm done. But when the inversion is caused by a large, slow-moving, high-pressure mass that lasts for days over congested, heavily-populated cities, the effect can be disastrous.

London's fogs are often considered a charming feature of the city, but they have long been evidence of enormous amounts of aerosols pumped into the air by industry, automobiles, and coal furnaces. Bad air has been a part of London life since at least the twelfth century, when coal was first used extensively for heating and industry. By the mid-1600s the situation in London was so dire at least one prominent citizen was prompted to ask King Charles II to do something about air so bad that he and his neighbors were forced to "breathe nothing but an impure and thick mist, accompanied by a fuliginous and filthy vapor, corrupting the lungs, so that catarrhs, coughs, and consumptions rage more in this one city, than in the whole Earth." Conditions grew even worse during the Industrial Revolution of the eighteenth and nineteenth centuries, but it was not until well into the twentieth century that Londoners were given sure proof that their careless fouling of the air could cause death.

On December 3, 1952, a layer of clouds formed high above London as warm air moved in and condensed to form the lid of a major inversion. During the next four days the city was clogged with a brown smog so dense visibility was reduced to as little as one foot. The city grew dark and cold and the natural response was to switch on lights and fire up coal stoves and electric heaters, adding even more to the already dangerous levels of sulfuric pollutants trapped beneath the inversion. Road and barge traffic came to a virtual standstill. People, especially the elderly, began to grow ill, filling hospitals. Then they began to die. By the time a low-pressure center moved in and winds dispersed the smog, more than 4,000 Londoners had died and thousands of others were hospitalized with respiratory ailments.

Soon after the disaster, legislation was passed that established new standards for smoke emissions, drastically reducing the amounts of pollutants. Within a few decades London's air was said to be 80 percent cleaner than before the 1952 inversion.

Frequent inversions in the Los Angeles area are caused by warm air masses that circle in over the Pacific, trapping ocean-cooled air below and creating the city's infamous smogs. Because Los Angeles is surrounded by the San Gabriel and San Bernardino mountains, the inversions tend to linger. When the Spanish explorer Juan Rodriguez Cabrillo visited the Los Angeles basin in 1542 he noticed that smoke from Indian campfires rose only a short distance in the air then spread horizontally—a classic symptom of an atmospheric inversion.

The Dew

IN EARLIER AGES, weatherwise farmers walking to the barn in the morning were apt to recall this old proverb:

> When the dew is on the grass,
> Rain will never come to pass.
> When the grass is dry at morning light,
> Look for rain before the night.

While the presence or absence of dew is not a surefire weather indicator, it does reflect what was going on in the atmosphere during the night. Heavy dew forms most often under clear skies, when the loss of radiant heat from the ground sets the stage for condensation. On the other hand, when there is no dew, it is often because a covering of clouds during the night prevented the ground from cooling.

Dew, like fog, is a product of moist air coming in contact with cool surfaces. At night, the surface of the earth cools, causing the warmer, moist air that comes near it to lose heat. Because cool air can not hold as much moisture as warm air, it becomes increasingly more saturated with water vapor until it reaches a temperature at which it can hold no more—its dew point. When there is no breeze to cause a vertical mixing of that cold air with the warm saturated air, condensation occurs not in the air (which would cause fog) but on any cool surface it touches. Thus molecules of water vapor coming in contact with a blade of grass condense into a droplet of water on the grass. That condensation combines with other moisture resulting from transpiration—the evaporation of moisture from the plant's cells—to produce the glistening dewdrops that soak our shoes when we walk across a lawn in the morning.

Because it seems to appear out of thin air and is only available in small quantities, dew was long thought to be a sort of nectar from heaven, with magical or rejuvenative properties. Washing your face in it, especially in the month of May, was thought to be good for the complexion, and drinking small amounts was as invigorating as cod-liver oil, and certainly tastier. During May Day festivities in Merry Olde England young maidens would sometimes roll naked in the early morning dew to make themselves more attractive. Even that dour old warrior Oliver Cromwell would now and then take a draught of May dew to give him the strength and good health of youth.

WinteR

Winter · solstice

23½°

INTRODUCTION

Winter is a season of contradictions. In many cultures, it is a time of feasting and gift-giving, when celebrations and goodwill give comfort against long nights and cold, wet, stormy weather. Christmas, the most important holiday of the year in the Christian tradition, is descended in part from very old pagan celebrations, the best known of which was the Saturnalia of ancient Rome. The Saturnalia was celebrated at the end of December with a three- to seven-day holiday in which all work and business ceased and slaves were free, for once, to do as they pleased. The Greeks much earlier had celebrated the Lesser, or Rural, Dionysia in late December, in honor of the god of wine and fertility, Dionysus, and the Halcyon days, in honor of the myth of King Ceyx and his wife Halcyone. Earlier yet, the Babylonians celebrated the Sacaea, a New Year holiday marked by feasting, drinking, sexual license, and mock battles reenacting the creation of the world.

Christmas also has some roots in the ancient Yule Girth festival of the Goths and Saxons. It was their custom on the night of the winter solstice to build huge bonfires on the tops of hills to help stoke the dwindling flames of the sun. Christmas revelers continue a remnant of that tradition by burning Yule logs and decorating their houses and Christmas trees with bright lights.

Although the season has frequently inspired generous and light-hearted celebrations, it has also been the cause of less cheerful gatherings. The word *winter* is Anglo-Saxon, probably derived from the same root as "wind" (the ancient Saxons called November "wint-monat," or "wind-month"). It is a cruel season of darkness and decline, the realm of ghosts and witches, when the living world descends into sleep, death, and decay. For the ancient Hindus and Chinese it was a time to worship ancestors and appease the spirits of the dead. Scandinavians and Celts celebrated the cult of the dead in winter, lighting bonfires on hilltops to represent the annual death and rebirth of the sun. In Nordic mythology, Woden (or Odin), the god of the dead, wanders the earth in winter. The Icelandic sagas prophesy that the world will end in the midst of a terrible winter. Iranian myths predict mankind will be destroyed by devastating seasons of rain, snow, and ice.

In the Northern Hemisphere, winter's official beginning is December 21 or 22, the winter solstice, the shortest day of the year, when the earth is tipped on its axis so that the North Pole reaches its farthest point from the sun and is in the midst of a long, perpetual night. On the morning of the winter solstice, the sun rises at its southernmost point of the year, following its lowest arc, and does not rise at all anyplace north of the Arctic Circle, 23½ degrees from the pole.

Winter's actual beginning varies, depending on location, perspective, and the whims of weather. The naturalist Hal Borland insisted that winter in New England began on the first full moon after the middle of November. In northern Europe it was traditional to count the first day of winter as St. Martin's day—October 28, November 1, or November 11—when the saint riding his white horse signified the coming of snow. Botanists consider the season to begin when the average daily temperature falls below 43 degrees Fahrenheit, the point at which plant growth ceases, sap stops flowing, and leaves, stems, and flowers wither and die.

In temperate regions the winter solstice, though it is the shortest day of the year, will probably not be the coldest day of winter. In much of the Northern Hemisphere, December 21 scarcely qualifies as a winter day, seldom as cold or snowy as the days of January and February yet to come. The lag is explained by the ability of the earth (and particularly the bodies of water that cover three-quarters of its surface) to absorb and hold heat, in much the way a fireplace brick remains warm long after the fire is out. It requires many weeks of cold weather before that stored heat is lost. Full winter may not arrive until well into January and may last even after the days have lengthened and should have warmed enough to melt snow and cause plants to sprout with new life.

Winter means different things in different parts of the world. In the American Deep South, midwinter is considered to occur around the winter solstice, and in exception to the rule farther north, temperatures do not normally continue downward after that date. In tropical and subtropical regions, rather than a season of snow and subzero temperatures, it is a season crowded by new-comers. In addition to thousands of human refugees from the north, those areas experience a dramatic influx of migrant birds. The crowding can be significant, especially where cutting of the rain forests has increased the competition for space.

Winter can come as early as late August in the far north, and may really take hold by September. North of 70 degrees North, the cabin-fever territory above the Arctic Circle, the sun disappears for good from November 16 to January 18; above 80 degrees North it is gone from October 19 to February 22.

The farther north you live, the more likely you are to suffer from the winter blues. Medical researchers, treating the condition as a mild form of depression, named it SAD—seasonal affective disor-der—and say as many as 25 percent of people living in northern latitudes suffer from it. The symptoms include chronic depression, fatigue, ennui, increased sleep, weight gain, and a desire to with-draw socially. One theory blames a hormone called melatonin for those symptoms, after discoveries that the level of that hormone in the blood increased dramatically as days shortened. Treatment with artificial lights bring melatonin levels back down, and in many cases reduces the symptoms of SAD.

None of the scientific data about winter—the observations and analysis, the computer-generated maps, the complex models de-signed to give us some hope of approaching spring—attracts as much interest as the appearance of a groundhog on February 2. According to popular wisdom, if the groundhog (or woodchuck) sees his shadow, we can expect at least six more weeks of winter, but if no shadow is visible, we can look forward to an early spring. The groundhog legend found its way to North America via Ger-many, where the weather on February 2—Candlemas Day—was of high interest to a populace who celebrated outdoors that day. In its original form, no animal was involved, but later a badger or bear was elevated to the status of weather prophet. There may be some meteorological validity to the folklore, since in February clear skies and sunshine—and thus groundhog shadows—often occur during cold, clear, and stable atmospheric conditions, an indication of major cold masses and long periods of cold weather to come.

Are Winters Getting Milder?

EVERYONE SEEMS TO remember a time when winters were harder than they are today. The standard argument against that observation is that memory is notoriously imperfect. We tend to remember the unusual and difficult, and besides, the snow that seemed so deep when we were children would hardly top the boots of an adult. Still, an apparent softening of winter in many parts of the world has been argued for decades and has in recent years been proposed as evidence of global warming caused by the greenhouse effect (though the scientists who support the global warming theory usually insist the effects will not become apparent until well into the twenty-first century). Meteorologists are more likely to explain changes in climate as normal fluctuations of varying durations.

Claims that winters are growing milder are nothing new. A Canadian journalist in 1853 wrote in *Colonial Magazine*: "It appears that Canada has already relaxed some of its former rigors, and is in a state of continued mitigation. Since a portion of its forests have been cleared, its swamps drained, its villages and settlements established, the Indians inform us that the frosts have been less severe and frequent—that the snows fall in smaller quantities, and dissolve sooner."

There is irrefutable evidence that climates fluctuate over time. Paintings of the Alps dating back to the Middle Ages show glaciers where there are none today, evidence for the "Little Ice Age" that lasted from about A.D. 1200 to the mid-1800s. Such climatic changes over relatively short periods of time could be caused by fluctuations in the sun's heat, alterations in the earth's orbit or angle of axis, or changes in the atmosphere.

It is possible that winters in some parts of the world were colder and snowfall amounts heavier not many decades ago. However, if you were to base your opinion only on cold temperatures in the United States, you might reach a different conclusion. In the first ninety years of the twentieth century, the coldest winter in the United States was not 1935–36 or 1948–49, as your grandfather might insist, but 1978–79. That year the mean winter temperature for the forty-eight contiguous states was 27 degrees Fahrenheit, several degrees colder than any other winter in the twentieth century, and more than five degrees colder than the eighty-nine-year average of 32.5 degrees.

Scintillate, Scintillate, Little Star

BECAUSE STARS TWINKLE—or scintillate—in the night sky it is natural to assume distant suns burn unevenly, like bonfires. Yet, viewed from above our atmosphere, stars do not twinkle at all. The pulsating lights we see are a visual distortion caused by layers of air of differing temperatures and turbulence in the atmosphere. The light that finally reaches our eyes is being constantly distorted and shifted by moving "blobs" of air. Get high enough off the ground and the distortion is minimized, which is why most observatories are located on mountain tops, above the layered air of low altitudes.

In some tropical countries the increased twinkling of stars high in the sky is considered a sign that the rainy season will soon begin. Greater than usual scintillation is caused by unstable air, increasing humidity, and high winds in the atmosphere, and is a reliable indication of approaching precipitation. Also, watch the colors of the stars: High humidity makes stars twinkle blue, further evidence of coming rain.

OWLET
MOTH

POLAR
BEAR~
Winter Den

PROMETHEA
MOTH~
Leaf
Cocoon

FROG~
In Hibernation

GOLDENROD
GALL

EVERGREEN
BAGWORM~
Bags

TURTLE~
In Hibernation

HONEY
BEES

BARK BEETLE GALLERIES

COLD

It is a cold occupation, watching for meteors in January. For relief we have built a campfire near the river, where the canoes are pulled up high on the bank and the tents are nested in the snow. We return now and then to huddle over the fire, kick a log into place, watch the sparks flush swirling toward the stars. Then we go back to stand in the darkness and watch for the brilliant pebbles flaring across the sky. We are invigorated and determined, but mostly we are cold. The cold penetrates the layers of our high-tech winter clothing, and drains the heat from our earlobes and fingers and toes—the dangerous frontiers far removed from the heat-pumping heart. Once started at a fingertip, you can feel cold creeping up the bone, like rust.

Because cold air can not hold as much water vapor as warm air, the atmosphere on cold nights can be at its cleanest and clearest. This night the stars are so bright and defined they appear almost within reach. But there is a price to pay for the view. I'm reminded that cold is the original and ultimate state of the universe, and I try to imagine being adrift in the deep cold of space.

My imagination fails me. We are accustomed to thinking of cold in terms of temperatures that can freeze water, but in space *molecules* freeze. Scientists believe no lower temperature is possible than absolute zero, −460 degrees Fahrenheit, the temperature at which molecular motion ceases. They have designated the temperature of absolute zero as 0 degrees Kelvin, and have calculated that the average temperature of the universe is 2.7 Kelvins, or −454.8 degrees Fahrenheit.

One reason life on earth is possible at all is that the atmosphere shields us from the cold of space. Still, some places on earth experience cold so extreme it seems to plummet straight down from the void, as if the atmospheric blanket had opened for a moment, sucking all heat from the surface and allowing free passage of absolute cold and darkness. Our planet's coldest spots are the poles, those end caps of the globe that receive the least direct sunlight during the year and that in the winter are hidden in total darkness.

Most of us know what it means to be in bone-chilling cold, to feel our sinuses expand, to breathe air so harsh it seems to sear the lungs. But cold in the north and south polar regions of the earth takes on a higher degree of reality. Stories are told of cold so severe it causes lungs to hemorrhage; teeth crack and fillings pop out like orange seeds.

The North Pole is a bitter and inhospitable place, where there is no land, only a perpetual ice cover over an ocean thousands of feet deep, yet it can seem mild compared to the interior of Antarctica and the South Pole. That barren, mountainous, ice- and snow-covered continent, larger than the United States and Western Europe combined, averages at least 35 degrees colder than the Arctic, and holds the record for the earth's coldest temperature: −129 degrees Fahrenheit set July 21, 1983, in Vostok, Antarctica. At the South Pole the highest temperature ever recorded was 5.5 degrees Fahrenheit, and the average year-round temperature is −57 degrees Fahrenheit.

Temperatures remain so low at the southern polar ice cap because very little short wave radiation from the sun reaches the earth's surface there, since most of it (about 80 percent) is reflected away by the perpetual covering of snow. Only for a brief period in November and December—midsummer in the Southern Hemisphere—does the ice and snow surface of the interior absorb slightly more radiant heat than it loses.

In his 1937 book, *The Worst Journey in the World*, the British explorer Apsley Cherry-Garrard, a survivor of several expeditions to the South Pole, described cold so extreme that perspiration froze in sheets to the inside of his clothing, and especially in his sleeping bag, where he was forced to sleep with his head covered, breathing his warm but damp air into the interior. It was so cold in the mornings while his clothes were still damp but not yet frozen that he had to be conscious of his position when first coming out of

the tent because in fifteen seconds the clothing was frozen solid
and he had to maintain that same position all day. "The temperature was −66 [degrees Fahrenheit] when we camped," he wrote,

> and we were already pretty badly iced up. . . . For me it was a
> very bad night: a succession of shivering fits which I was quite
> unable to stop, and which took possession of my body for many
> minutes at a time until I thought my back would break, such was
> the strain placed upon it. They talk of chattering teeth: but when
> your body chatters you may call yourself cold. I can only com-
> pare the strain to that which I have been unfortunate enough to
> see in a case of lock-jaw. The minimum temperature that night
> as taken under the sledge was −69; and as taken on the sledge
> was −75 . . .
>   Our breath crackled as it froze. There was no unnecessary con-
> versation: I don't know why our tongues never got frozen, but
> all my teeth, the nerves of which had been killed, split to pieces.

Another explorer, Paul Siple, who accompanied Richard E. Byrd
on his first Antarctic expedition in 1929, later described some effects of the cold: "Vapor from a man's breath could freeze his
eyelashes shut in an instant and make him believe he had gone
blind. His breath would come in gasps and his joints would ache.
The intense pain of the cold on fingers and toes could easily dis-
tract him, and even destroy his ability to reason clearly."

In temperate regions, less extreme cold is an integral part of
winter. Temperatures of −40 or −50 degrees have been recorded
at many places in North America, where cold waves sweep down
from Arctic regions with all the rampaging regularity of Medieval
Turks. Such temperatures as the −81 degrees Fahrenheit that Snag,
Yukon Territory, experienced February 3, 1947, and the −70 degrees that was recorded at Rogers Pass, Montana, on January 20,
1954, appear to be possible only when skies are clear (clouds insulate the earth and hold heat in), and atmospheric conditions are
dry and calm. The coldest temperatures are usually recorded as
well at points far above sea level, in basinlike valleys where cold
settles to the bottom.

Surface temperatures on earth depend ultimately on sunlight. In

April 1815, the volcano Tambora in the Dutch East Indies erupted, spewing massive volumes of ash and gas into the atmosphere. The following year a dense cloud of ash obscured the sun in Europe and America, and probably caused the extensive crop failures and severe famines throughout that entire "year without a summer." It may have been responsible as well for the severe cold weather that struck New England the following winter (1816). That winter became known by New England farmers as "eighteen-hundred-and-froze-to-death."

It is because of cold that bears den, wrens migrate, and frogs burrow in the bottoms of ponds. Grouse sometimes roost in loose snow to avoid cold, and songbirds huddle, shivering, in shrubbery to find protection from it.

Insects have evolved some of the most ingenious tactics in the animal world to deal with the challenge of cold temperatures. The harshest, most pragmatic adaptation is to spend a summer feeding and finally mating, lay eggs in protected places before the cold weather begins, then die, as if in acquiescence to the insurmountable difficulties of winter. Less drastic are the tactics of many aquatic insects, which burrow into lake and river bottoms, and those that live in the soil, which keep from freezing by burrowing deeper into mud and soil. Many beetles and flies take cover in tree bark or under leaf litter and enter a state of diapause, or suspended animation. Most moths overwinter as eggs or pupae, but the owlet moth (one of 101 species of *Cucullinae* found in New England) passes the winter in adult form by taking cover under leaves that are themselves covered over with snow; they mate and lay eggs in early spring, about the time tree buds break out and shortly before most migrant birds return to prey on them. The larvae of the Khapra beetle of India (*Trogoderma granarium*), will enter diapause during low temperatures, as well as when threatened by drought or starvation, and can stay in that state for as long as eight years. In a tactic known as "balling," honeybees cluster in dense masses and activate their wing muscles to give off heat, maintaining a constant hive temperature well above freezing. Other insects have adapted defenses that allow them either to resist freezing or to survive it.

Many insects are able to overwinter because they "cold-harden" themselves in a resting stage during which they empty their guts of fluids or produce various fats and sugars in their body fluids so that they will only freeze at temperatures well below the freezing point of water. Some insects, providing they remain motionless, can keep the water inside their bodies supercooled to resist freezing in temperatures as low as −30 degrees Fahrenheit. Others, such as caterpillars of the gall moth and a scavenger beetle found in subarctic regions, can survive virtually any cold weather by synthesizing into their body fluids a 40 percent solution of glycerol— the same ingredient used in automobile antifreezes. When fully equipped with this antifreeze the caterpillars and beetles can survive (for a short time, at least) in temperatures as low as −125 degrees Fahrenheit.

The hardiest of insects are those that simply let themselves freeze. In the arctic, the woolly bear caterpillar, *Gynaephora*, spends ten months of each year frozen at temperatures as low as −60 degrees, thawing out and growing only during the brief summers, until after as long as thirteen years it reaches maturity and can metamorphose into a moth. Such indifference to freezing is possible only because of a high degree of cold-hardening that allows the insect to control and slow the freezing process, allowing body cells time to adjust to the stresses of freezing.

Our own body cells, unfortunately, are not so adaptable. We humans are pathetically ill-equipped for extremes of temperature. As the night progresses I spend less time looking for meteors, more time getting acquainted with the fire. At some point, late, a flash of light catches my attention and I look up to see a meteor scorch a line across the sky, like a match drawn across sandpaper. I wait, but not a trace of heat filters down.

Chilly Winds

COLD IS MAGNIFIED by wind. In fact, temperatures that might be considered only moderately cold can be made uncomfortable, and even dangerous, by the windchill effect of wind. Cold air blown by wind increases convective heat loss at a much faster rate than cold air alone. Thus, your chances of being frostbitten are better in a forty-mile-per-hour wind at 30 degrees Fahrenheit, than in calm air at −40 degrees. In extremely low temperatures, a breath of breeze can make the difference between an animal's survival and death.

To figure the windchill factor, multiply the wind speed by 1.5, and subtract it from the air temperature. For example, if the air temperature is 20 degrees Fahrenheit and the wind is blowing at 30 miles per hour, use the formula 20 − (30 x 1.5) to determine that the windchill factor is −25 degrees Fahrenheit.

air temperature
− windspeed x 1.5
windchill

Stars with Horrid Hair: Comets

... and like a comet burned,
That fires the length of Ophiuchus huge
In the arctic sky, and from his horrid hair
Shakes pestilence and war.
—John Milton, *Paradise Lost*

Returning from the edges of the solar system, from distances so immense they sometimes penetrate the black and empty regions between the sun and the nearest star, alpha Centauri, comets have always blazed a vivid trail through the human imagination. In eras when it was widely believed that the future could be read from the regular, chartable, dependable motions of the stars and planets, comets represented the unknown and mysterious. They appeared without warning, grew brighter and brighter, then dimmed and disappeared. They were unpredictable. They seemed to suggest powers we knew nothing about.

Like all anomalies of the skies, comets through most of human history have been viewed with alarm and suspicion, and as portents of ill tidings. In ancient woodcuts and tapestries they were often represented as swords or daggers hanging in the sky, or as decapitated heads with the hair streaming behind them, a particularly telling image since the Greek root of the word comet, *kometes*, means "hairy," in reference to a comet's wispy tail. A comet that appeared in 44 B.C. was interpreted by Romans as a harbinger of the assassination later that same year of Julius Caesar. Another that appeared in A.D. 837 was thought to presage the death of Roman emperor Ludwig the Pious in A.D. 840. When the hairy star that

WHIPPLE'S DIRTY SNOWBALL

JET

CRUST

solar wind

PERIHELION

would centuries later become known as Halley's Comet made an appearance over Jerusalem in A.D. 66, despairing residents predicted that their city would fall to the Romans (it did, four years later).

Until the late 1600s, the era of Isaac Newton and Edmond Halley, almost nothing was known about the nature and origin of comets. The Danish astronomer Tycho Brahe had been among the first to observe them as astronomical rather than supernatural, divine, or atmospheric phenomena, calculating that they must be at least six times farther from the earth than the moon and thus disproving Aristotle's long-accepted notion that comets were flames burning slowly in the upper atmosphere. Yet even Tycho believed that the comet he observed in 1577 would "spew its venom" across the earth and cause war, disease, and religious strife. From the brief observations of a few visible comets it seemed impossible to determine whether they orbited the sun like the planets or passed through the solar system in a straight line. Even the most careful astronomers came to believe that comets wandered through space as random phenomena and that once they passed from view they would never appear again. Thus Johannes Kepler, who was correct in asserting early in the seventeenth century that comets shine with reflected sunlight and their tails are spread in an antisolar direction by solar radiation, would cling to the mistaken notions that comets traveled in straight lines rather than orbits, that they were created spontaneously in the ether from "fatty globules," and that they were guided through space by an innate intelligence.

From his observations of the comet of 1680–81, Isaac Newton calculated that the comet's trajectory was a parabolic orbit of the sun. His friend Edmond Halley would use that idea and the mathematics pioneered by Newton to demonstrate, for the first time, that comets can make more than one appearance.

Halley's name became linked with the most famous of periodic comets following his observations of it in 1682. After collecting all the information he could find about previous comets he noticed that the one studied by Johannes Kepler in 1607 had passed through the same portion of the sky as the 1682 comet. Digging

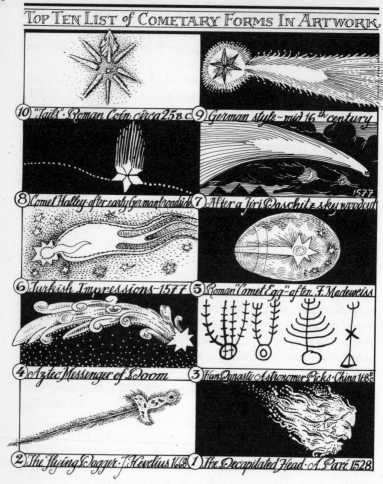

Top Ten List *of* Cometary Forms In Artwork

10) "Tails" Roman Coin, circa 25 B.C. 9) German style ~ mid 16th century

8) Comet Halley after early German Broadside 7) After a Jiri Daschitz sky woodcut 1577

6) Turkish Impressions ~ 1577 5) Roman "Comet Egg" after F. Madeweiss

4) Aztec Messenger of Doom 3) Han Dynasty Astronomer Picks China 168 B.C.

2) The Flying Dagger · J. Hevelius 1668 1) The Decapitated Head · A. Paré 1528

deeper Halley discovered that comets witnessed in 1456 and 1531 had also passed through that same region of the sky. He concluded that these sightings, occurring at intervals of seventy-five or seventy-six years, were of the same comet, and that it must be traveling an enormous, elliptical orbit. If his calculations were correct, he wrote, another comet should appear in 1758.

Halley did not live to see his prediction come true, but from the moment the returning comet was first seen on Christmas Day 1758 his name was permanently attached to it. The recorded history of Halley's past returns goes back to 240 B.C., when it was seen in China and blamed for the death of the Empress Dowager. The A.D. 684 return was recorded in the *Nuremberg Chronicle*, with the usual list of resultant pestilences, failed harvests, and destructive weather. It appeared in 1066, the year of the Norman Conquest, and was popularly believed to have presaged the defeat of Harold of England by William the Conqueror. The Bayeux Tapestry, which commemorated the Norman victory with seventy scenes, shows Halley's Comet as a prominent star with a multiple tail, being pointed at by concerned men on the ground.

The return of Halley's in 1910 was accompanied by intense excitement and a bit of hysteria. It was only the third return since Edmond Halley worked out the comet's periodicity, but already much more was known about comets in general and Halley's in particular. The moment of its perihelion—the point in its orbit when it would be closest to the sun—was calculated to within three days, and scientists had a pretty good idea what to expect. Still, there was widespread public misunderstanding and mistrust of the event. The announcement that the earth's orbit would carry our planet through the tail of the comet convinced many people that the end of the world was near. A rumor spread that the gases in the tail were lethal, and profiteers were inspired to produce and sell "comet pills" they claimed to be antidotes to the comet's gases.

Halley's Comet is sometimes called "the comet of all lifetimes," or "mankind's comet," because the length of its orbit brings it within sight of the earth about every seventy-five years—the approximate length of one human lifetime. Most people who live to adulthood have the opportunity to view it.

A theory of the structure of comets proposed by American astronomer Fred Lawrence Whipple in 1950 is still accepted by most scientists as accurate. According to Whipple, a comet is essentially a giant, dirty snowball, composed of frozen water, ammonia, methane, carbon dioxide, hydrogen cyanide, and other compounds, and embedded with particles of dust and rock. Like asteroids, comets were probably left over after the formation of the solar system, when clouds of gas and dust contracted to form the sun and the planets. Also in 1950, the Dutch astronomer Jan Oort hypothesized the existence of a swarm or "reservoir" of comets located far beyond Pluto's orbit. Swirling within the enormous, spherical "Oort Cloud," one to two light-years from the sun, could be as many as 100 billion comets. Occasionally, collisions within the cloud propel a comet close enough to the inner solar system to be captured by gravity and brought into an orbit within sight of earth. Oort's theory explained why new comets periodically appear to replace those that have disintegrated and disappeared after repeated revolutions near the sun.

ORBIT OF COMET HALLEY

NEPTUNE

HALLEY'S COMET · 1066 · after the BAYEUX TAPESTRY

16 KILOMETERS

NUCLEUS

Whipple, in propounding his "dirty snowball" theory, described the core of a comet as measuring a few miles across, composed either of solid rock or solid ice. While the comet is in the outer end of its orbit it remains frozen solid and is so far away and reflects so little light that it is invisible from earth. Approaching the sun, however, it heats up, causing some of the ice to evaporate and spew gases, dust, and other particles. The nucleus remains solid, but becomes surrounded with a vast "coma" of dust as broad as the diameter of a planet and sweeping away in a tail that might stretch millions of miles. Once it passes the sun at the nearest point, or perihelion, of its orbit, the comet cools, the coma disappears, the tail shrinks, and the comet becomes less and less visible. Others, like Halley's, take longer to complete an orbit and thus can be expected to remain spectacular for a longer time.

When Halley's Comet returned in 1985–86 it was met by two *Vega* and one *Giotto* spacecraft from the USSR and Europe, which passed near the comet's head and took photographs. The photos showed a somewhat hilly, porous, ellipsoidal nucleus measuring five miles by ten miles. It was black in color, but composed of ice (as Fred Whipple had predicted) made up of 80 percent water, 17 percent carbon monoxide, and 3 percent carbon dioxide. It traveled through space like a poorly thrown football, spiraling one revolution every 7.4 days at the same time it tumbled end over end once every 52.2 hours, all the while jetting up to seven enormous whalelike spouts that ejected three tons of dust and twenty-one to sixty tons of gas and water vapor every second.

Comets are classified as either "short-period" or "long-period." Short-period comets complete an orbit in less than 200 years; long-period orbits in more than 200 years. Short-period comets, like Encke's, with its three-year, six-month orbit, make frequent passes near the sun and disintegrate fairly rapidly, resulting in dim comets. Biela's Comet, with an orbital period of six years, nine months, was closely observed several times in the nineteenth century, but when it appeared in February 1846 it had split into two halves, each with its own tail. When it returned in 1852 the two halves had separated, one slightly ahead of the other, both sprewing long trails. It was never seen again.

Long-period comets have such long orbits it is difficult to anticipate them. Occasionally a "new" comet appears, one that might
last have been visible from earth long before the first humans were
here to see it. The Comet of 1811 was visible for eighteen months,
was extremely bright week after week, and at one point had a tail
more than 100 million miles long—longer than the distance from
earth to the sun. Another, in 1843, had a tail that extended across
about one-fourth of the sky. It was a class of comet known as a
"sun-grazer" because it passed so close to the sun—about 80,000
miles from the surface, as opposed to Halley's 55 million miles.
Such proximity to the sun caused the comet to accelerate to a
speed estimated at 1,270,000 miles per hour, fast enough to propel
the comet most of the way around the sun in one day, like an ice-
skater "cracking the whip," before being shot back out toward the
outer reaches of the solar system. Donati's Comet, in 1858, had
several irregular tails that slowly changed shape, giving support to
the theory that gases inside a comet build pressure and periodically
explode, causing spurts of particles to issue from the nucleus. The
Comet of 1861 was particularly spectacular because it approached
to within 11 million miles of earth, half the distance to Venus, and
its tail actually swept across our planet. Some witnesses later
claimed the Comet of 1861 was a warning from God of the coming
horrors of the Civil War. The Comet of 1882, a sun-grazer that
followed nearly the same orbit as the Comet of 1843, split into two
fragments, the smaller of which gradually disappeared from view
while the main body continued on its path.

Ancient fears of comets may not be entirely unfounded. One
respected theory blames the extinction of the dinosaurs (and perhaps other periodic mass extinctions) on the impact of a comet,
asteroid, or other large body. If a comet the size of Halley's struck
our planet, we could expect to experience a mass extinction first-
hand. Some scientists believe a fragment of a comet much smaller
than Halley's may have been responsible for the so-called Tun-
guska Event of 1908, when an enormous blast leveled 750 square
miles of forest in central Siberia. Witnesses reported seeing a large
and elongated object flaming across the sky moments before a

blast so immense that shock waves were felt around the world and a cloud of dust obscured the atmosphere for days afterward in London—6,000 miles away. Trees devastated by the impact all laid pointing away from a central point, yet there was no crater, leading to the theory that the explosion was caused by the shock wave of a comet or asteroid vaporizing in the atmosphere.

Of the dozen or so comets that appear each year about half are new and unpredicted. Some of them are first spotted by amateurs with backyard telescopes. A few become visible with the naked eye but most of the visible comets are so distant and appear so small it is essential to know where and when to look for them. Two good sources of up-to-date reports are *Sky and Telescope* magazine and the "Celestial Events" column of *Natural History* magazine.

Few comets with long, spectacular tails have been visible in the last half of the twentieth century, but that situation could change any time. For determined skywatchers, patience is definitely a virtue. If no other comets appear first, you can always wait for Halley's: It is due again in 2061.

COSMIC LITTER

I remember as a child standing outside at night watching stars and being awestruck by a single, dim object moving soundlessly across the sky. My father told me it was a satellite—an artificial moon, constructed and launched into orbit by men in Florida or the Soviet Union—and that even as we watched it was radioing images and information back to the ground. The idea that something human-made could be up there with the stars seemed astonishing and frightening to me.

Now, thirty years later, it is possible to stand outside for an hour on a dark night and easily spot a half-dozen satellites. They come and go routinely, like aircraft in a flight pattern over La Guardia. There are hundreds in orbit around our planet, busily receiving and transmitting information to meteorologists, astronomers, communications companies, and the military. Since the first *Sputnik* was launched on October 4, 1957, more than 3,600 satellites, probes, and manned stations have been sent into orbit. Space, at least our corner of it, is getting rather crowded.

Those crowds of orbiting objects give off clues to their identity. Most civilian satellites travel west to east (never east to west), while military surveillance satellites more often travel north to south or south to north. All vary in brightness according to their size and altitude, and sometimes appear to be faintly blinking, as if tumbling and rhythmically exposing a side that is more reflective than the others. If they pass into the sunless umbra on the back of the earth they fade from view and disappear, only to reappear a few moments later as they pass into the light again. The speed of a satellite's orbit varies with its distance from earth. At a height of 200 miles, it orbits once every 90 minutes; at 500 miles it orbits in 100 minutes; at 22,300 miles it orbits in 24 hours, synchronous with the time it takes the earth to rotate once on its axis.

All satellites have a finite life. Some are ultimately rescued and brought back by astronauts in spacecraft, others wear out and are left to orbit indefinitely. A few fall to earth. Years ago, as friends and I drove on a highway along the Lake Michigan shoreline, we spotted a large fireball blazing across the sky over the lake. We pulled to the side of the road and jumped from the car to watch. It came from the south, traveling too slowly for a meteor, and for a terrible moment we thought it might be a jet in flames. Yet there was no sound, the object did not seem to be losing altitude, and something about its velocity suggested it was much too high in the air to be an aircraft. We decided it had to be some extraordinary meteor, perhaps one grazing horizontally through the atmosphere before passing into space again. In perhaps a minute it had crossed the sky and disappeared over the horizon to the north. We learned the next day it had been a Soviet satellite burning up during its reentry into the atmosphere. It crashed in a remote area of northern Ontario, hundreds of miles north of us.

That was not the first piece of space junk, nor will it be the last, to come crashing to earth. An unmanned Soviet spacecraft burned up in the atmosphere over Canada in 1978, scattering radioactive debris from its nuclear reactor near the town of Fort Reliance in the Northwest Territories. *Skylab*, the seventy-six-ton vessel that was the first U.S. space station, fell out of orbit after 34,000 revolutions of the earth in July 1979, scattering debris over parts of Australia and the Indian Ocean. A five-ton Soviet spy satellite, *Cosmos 1900*, burned up in the atmosphere in October 1988. In November 1990, the third stage of the Soviet rocket, *Proton*, burned up over northern France a month after launching a communications satellite into orbit.

In addition to dead satellites and discarded rocket stages, there have been thousands of pieces of litter left circling the earth in the wake of the space programs. In 1961, less than four years after the Soviet satellite *Sputnik I* initiated the space age, the U.S. Air Force was tracking about sixty discarded objects in orbit around the earth. That same year a rocket used to launch a U.S. satellite exploded into about 300 fragments. By 1966 the Air Force reported at least 1,300 substantial pieces of orbiting litter. As of late 1991, at least 7,000 objects more than four inches in diameter, and nearly 150,000 objects larger than half an inch in diameter could be tracked with radar. Thousands—perhaps millions—of other objects are too small to be detected with radar, yet are significant enough to create light pollution as sunlight glints off them and interferes with astronomers' observations.

Space junk orbits at speeds of 9,300 to 22,500 miles per hour, fast enough to lend a ball of aluminum half-an-inch in diameter the impact of a rifle bullet. A .008-inch fleck of paint—about the size of a grain of salt—that struck the windshield of a U.S. space shuttle in June 1983 left a sizable pit in the glass and caused engineers to realize how hazardous larger projectiles could be. Since then seventeen other space shuttle windows costing $50,000 each have been replaced after being damaged by such impacts. A bus-sized satellite that spent seven years in orbit and was captured and brought back to earth by the space shuttle Columbia in January 1990 was found to be pitted with thousands of minute craters. After analysis, scientists concluded that less than half the craters were caused by natural meteoroids. The majority were the result of collisions with human-made debris.

Orbiting around our planet are about 3,000 tons of rubbish, including thousands of fragments of satellites left behind after the testing of U.S. and Soviet antisatellite weapons, plus rocket casings, solar panels, protective shields, clamps, straps, nuts and bolts, lens covers, and other miscellaneous hardware. Most of the large debris, and countless flecks of paint, bits of ceramic, and tiny fragments of various metals, orbit the earth at the outer fringes of the atmosphere, 300 to 500 miles above earth. Other debris exists 14,000 to 22,000 miles away, in a region known as the geostationary orbital band. Many of the most recent satellites, especially those used for communications, are sent to those distant regions where they remain in geostationary orbit, rotating at the same rate as the earth and thus maintaining the same relative position, usually somewhere over the equator.

What goes up usually comes down, but in the case of space litter orbiting thousands of miles above the earth, it may takes centuries. Objects in lower orbits reenter the atmosphere and are incinerated much sooner. At least a few of the meteors visible from earth are these burning bits of human-made debris. Space junk usually travels much slower than natural meteors—2.5 to 5 miles per second, as opposed to 9.3 to 43 miles per second—and often breaks into fragments that glow with green or blue light as their magnesium and aluminum burn.

The problem of cosmic litter is being taken seriously by scientists —seriously enough to inspire international efforts to reduce the amount of debris left behind in space. Nonetheless, the U.S. National Aeronautics and Space Administration recently reported that the amount of junk in orbit around our planet could double by the year 2010. It is a situation not unlike the one people faced when we realized we could not keep dumping our wastes into rivers and oceans. Out of sight, out of mind? For a short time, maybe, but junk has a way of coming back to haunt us.

FLOWERS OF ICE: FROST

That old fellow Jack Frost may nip at our noses and ears, and in extreme cases cause painful and dangerous frost bite, but he is a kindly favorite uncle compared to the frost giants of Scandinavian mythology. They are the harbingers of not only frost and cold, but the dark, deep winter of doom that will precede the end of the world.

Interestingly, frost has inspired more mythological invention than either snow or ice, perhaps because of its complexity and elaborate beauty. In a Japanese legend the Frost Man is the slightly malevolent brother of the Mist Man. In Finland and northern Russia, a Frost Woman and a Frost Man were important deities with the power to control the weather and were regularly placated with sacrifices to ensure they moderated the severity of blizzards and were not unduly hard on the reindeer herds. The Mordvins of eastern Russia set out bowls of porridge for the Frost Man to protect the next spring's sowing of crops. Elsewhere in Russia, Father Frost was a mighty blacksmith who bound the earth and its waters with chains of ice. In German folklore Old Mother Frost would shake the feathers from her bed and they would fall to earth as snow. An Australian aboriginal myth describes frost originating with the seven stars of the Pleiades, which according to the myth were once seven sisters who lived on earth but were so cold they sparkled with icicles and could not be seduced into living with men. They flew up into the sky to take refuge, and once each year ever since have pulled icicles from their bodies and thrown them down to earth. Jack Frost himself is a spritely, elflike character credited in nursery rhymes and folktales with coloring autumn leaves and tracing icy patterns on windowpanes. He probably originated as a relative of those fierce frost giants of Scandinavia. In Norse mythology, Kari, god of the winds, was father of Jokul ("ici-

HOAR
FROST

RIME

cle") or Frosti ("frost"). Jokul Frosti became, in the English tradition, Jack Frost.

To a farmer, frost on the pumpkin means the growing season has ended and the harvest, what remains of it, will soon give way to winter. In England frost is a generic term for freezing temperature, and "degrees of frost" refers to the number of degrees the temperature dips below freezing, but for most of the English-speaking world frost is an accumulation of ice crystals that occurs when water vapor or water droplets come in contact with freezing surfaces. Because cool air is heavier than warm air, it settles to the ground on cold, still nights, building up a layer that may be many degrees cooler than the air only a few feet above it. When land surfaces are cooled below 32 degrees Fahrenheit, the water vapor contained in the air surrounding it condenses directly into the light, feathery deposits of ice we know as frost, which can adhere to any cold surface. Once enough cold air has accumulated it begins to move, flowing slowly downhill, filling every hollow, gulch, and valley, and creating frost on every tree, shrub, plant, rock, windowpane, roof, or other object with a surface temperature below freezing. Cold air, like any heavy, flowing substance, will continue to overflow, drifting downward into other hollows and valleys, and flowing ultimately, if time and temperature allow, to the sea. Even during the hardest frosts, the freezing air is unlikely to rise more than about ten feet above the ground, which is why a garden or orchard set on a hill is much less likely to suffer damaging frosts than one set in a valley or on a plain. Planting an orchard in a lowland surrounded by higher ground—sometimes called a "frost hole"—is an invitation to trouble.

Farmers long ago learned some tricks for predicting frosts. "Clear moon, frost soon," refers to the tendency for the earth to cool faster on clear nights. If the moon's features are dulled, fuzzy, or otherwise obscured, it is evidence of an approaching warm front, and frost, of course, would be unlikely. Farmers might notice also that evenings when the wind was still and the sky clear, with surface temperature around 40 degrees, are often followed by killing frosts. Evenings with colder temperatures, if they are accompanied by wind or cloud cover, are seldom followed by frost.

If farmers can anticipate when frost is likely, there are a number of tactics that might save their crops. Small gardens can be covered with a screen of cloth to trap radiant heat inside and prevent contact with moisture-laden air outside. Orchardists build bonfires or ignite smudge pots at intervals among their trees or use enormous fans to circulate air among their fruit trees to prevent cold air from settling around them. In extreme cases, spraying flowers and vegetables with water covers them with an insulating layer of ice that actually protects plant tissue from freezing.

There are two basic types of frost. The most familiar is *hoar frost*, the frozen equivalent of dew. It is caused when water vapor in still, humid air sublimates directly into ice crystals on a frozen surface. Sublimation can be either water vapor changing to ice or ice changing to water vapor, but in either case it skips the usual intermediate step of becoming water. When the air is extremely cold and the air heavily saturated with water vapor, hoar frost develops into distinctive, delicate, and feathery crystalline shapes. *Rime* forms when fog or low-hanging clouds containing supercooled droplets of water come in contact with an object and freezes there. It grows in the direction of the wind, as more droplets come in contact with the frozen crystals, and builds feathery or needlelike formations, often projecting straight out on the windward side of three trunks, telephone poles, fence posts, and other objects. The stronger the wind and the larger the supercooled droplets, the more extensive the rime formations will become. Formations up to three feet long have been noted on mountain tops.

The Dreaded Pogonip

IN WINTER, when dense fog forms on the eastern slopes of the Sierra Nevada Mountains in northern Nevada, it sometimes freezes into rimelike ice crystals known locally as the *pogonip*, a Shoshone

word for "white death." As the ice crystals drift on the air they promptly adhere to trees, bushes, grass, fences, telephone poles, buildings, rocks, or any other surface they come in contact with, forming spectacular displays of frost-flowers and ice-blossoms. Some witnesses claim the crystals drifting in the air can be so dense they make a lyrical "tingling" sound as they collide. Others have insisted those same ice crystals can cause pneumonia and other ailments. In 1887 the *American Meteorological Journal* reported that mountaineers called the pogonip "a sort of frozen fog that appears sometimes in winter, even on the clearest and brightest of days. In an instant the air is filled with floating needles of ice. To breathe the pogonip is death to the lungs. When it comes, people rush to cover. The Indians dread it as much as the whites. It appears to be caused by the sudden freezing in the air of the moisture which collects about the summits of the high peaks."

Scientific American printed a similar report in 1892, mentioning that "The pogonip fog is peculiar to elevated altitudes in the Nevada Sierras. It ascends from the valleys, and its chill embrace is so much feared by the Indians, who are predisposed to affections of the lungs, that they change their camp if apprised by the atmospheric conditions that the dreaded fog is approaching."

Witnesses in more recent times have been happy to report that the pogonip is harmless. The old fears may have been exacerbated by the prevalence of tuberculosis in the late nineteenth century, and the fact that breathing harsh cold air—whether it contained "needles" of ice or not—probably made existing lung conditions seem worse.

Nature Baroque: Snowflakes and Crystals

What child can resist tasting snowflakes? Head back, mouth open, tongue stretched to catch that bit of coldness—for children, snowflakes on the tongue are as essential to a northern winter as sleds and snowmen. We take it for granted that snow will fall, that every flake, its constituent crystals arranged in elaborate symmetry, is a wondrous and unique creation. But is it? And how are they formed? What chemistry is at work in those dense, dark clouds of winter?

There is more to the birth of a snowflake than Aristotle's assertion that "when a cloud freezes there is snow." Snow is not merely frozen rain. Rain occasionally freezes, falling to the ground as sleet or freezing rain, but snow originates independent of atmospheric drops of water. Individual ice crystals form high in the atmosphere when water vapor freezes around dust or other particulates. Without particles to serve as condensation nuclei, water vapor can be cooled to −40 degrees Fahrenheit before freezing occurs. A supercooled cloud of this sort seeded with a few particles often escalates into a snowstorm. The individual crystals collect additional molecules of water vapor one at a time, building on one another symmetrically in a rapidly growing, widening circle. Temperature, wind, humidity, and even barometric pressure will determine the growth and ultimate form of the crystal. Large and elaborate crystals form at higher temperatures and humidity, while the small, basic crystals such as those common in polar regions form when temperature and humidity are very low. As the crystals fall they bump against each other, breaking off pieces of ice that in turn serve as nuclei for new crystals. As they pass through warmer layers of air they adhere to one another, congregating into snowflakes that may contain a thousand or more crystals.

Snowflakes, then, are aggregates of snow crystals. When the temperature is near or slightly above freezing, snowflakes become wet, adhere to other flakes, and grow to two or three inches in diameter. On very rare occasions they can grow larger yet. According to a report in a 1915 issue of *Monthly Weather Review*, a snowfall on January 28, 1887, dropped flakes "larger than milk pans," measuring fifteen inches in diameter by eight inches thick across several square miles near Fort Keogh, Montana.

Only when the temperature remains consistently below freezing will complete, individual crystals fall to the ground. If the temperature of the cloud they form in and air they descend through is warmer than 27 degrees Fahrenheit, the crystals tend to be flat and hexagonal. Between 27 and 23 degrees, they tend to be needle-shaped. Between 23 and 18 they are likely to be hollow and columnar, with prismatic sides. At temperatures below 18 they can be columnar, hexagonal, or fernlike. Virtually all have six sides. That hexagonal tendency is something of a mystery, although some scientists suggest it is produced by electrical charges in the crystals, while others say it is basic to the molecular structure of water molecules. The atoms in a molecule of H_2O are arranged, in physicist Hans C. von Baeyer's graphic description, "with two little hydrogens stuck onto a big oxygen like ears on Mickey Mouse's head." Scientists like von Baeyer believe the angle at which hydrogens protrude from the oxygen atom—about 120 degrees—causes snow crystals to grow to a six-pointed symmetry that repeats the molecular structure of water.

The scientifically untestable notion that no two snow crystals are alike is probably true. There are too many variables involved: Did the crystal form around a nucleus of volcanic ash, or a bit of sea salt, or a fleck of industrial waste? At what altitude did it form? What temperature was the air it passed through? How much humidity did it contain? For two crystals to be identical they would have to form in exactly the same conditions, collect the same number of molecules of water vapor in the same order, and bump the same number of other crystals during their long descent to the ground.

No two snow crystals may be identical, but general categories or types have been identified for years. Beginning in the 1880s, a Vermont farmer and amateur photographer named Wilson Bentley began examining ice crystals and photographing them under magnification. Armed with apparently unlimited patience, a microscope, and a box camera, Bentley produced over 6,000 photographs, 2,000 of which appeared in his 1931 book *Snow Crystals*, and identified hundreds of types of crystals.

Serious study of snow crystals was performed in 1910 by a Russian meteorologist who identified 246 types in 176 days of observation. In the 1930s Japanese meteorologist Ukichiro Nakaya consolidated the list to seventy-nine categories of crystals plus anomalies and oddballs he called "mavericks."

In 1951 the International Commission on Snow and Ice simplified things immensely by devising a classification system recognizing seven basic forms of snow crystals: plate crystal, stellar crystal, column, needle, spatial dendrite, capped column, and irregular crystals.

Stellar crystals (or simply "stars") are the classic, most familiar form of snow crystals, and the basis of the "no two alike" myth. They are not as common as aggregate flakes, irregular crystals, or asymmetrical crystals but we are familiar with them because of countless artists' renditions.

In addition to the basic crystals, snow can form into ice pellets when it is buffeted by strong winds that break off the points of the crystals and pack them into tiny balls. *Graupel* is formed by crystals falling through layers of supercooled droplets of water vapor, which remain liquid as long as they are suspended in the air, but freeze the instant they come in contact with anything solid and coat it in a dense covering of rime frost.

In places where snow is a frequent companion, it is sometimes personified as a living thing. In Japanese folktales it is Yuki-onne, the Snow Woman, who appears before men wandering in snowstorms and lures them to sleep and death. In Nordic mythology snow is the Old Man, an aged king of Finland named Snaer, whose daughters are Thick Snow, Thin Snow, and Snowstorm. To the Inuits of the far north snow appears in so many forms and shapes it requires an advanced vocabulary to describe it. To them, *api* is snow not yet touched by wind; *upsik* is snow changed by wind into a firm mass; *siqoq* is smokey snow blowing along the surface of the ground; *annui* is falling snow; *qali* is snow that sticks to the branches of trees; *saluma roaq* is a snow surface of very smooth and fine particles; *natatgonaq* is a snow surface of rough and coarse particles; and *det-thlok* is snow so deep snowshoes are required to walk in it. Dozens of variations—as many as 200, by some accounts—make it possible for Inuits and Eskimos to speak more precisely about snow than anyone on earth.

The winter vocabulary of the English language is growing. We have adopted the Russian word *sastrugi* to describe windblown drifts, common in the Arctic and Antarctica, that look like waves on water. Cross-country and alpine skiers have adapted a litany of descriptive slang expressions to identify the conditions they encounter on their skis. Among them are such colorful terms as *windslab, glop, fluff, névé, breakable crust, crud, sugar, corn, boilerplate,* and *cement.*

Explorers in Antarctica found to their dismay that in extremely cold temperatures (− 50 is not uncommon) snow can become unskiable. At those temperatures the tiny ice crystals that fall almost continuously, even from clear skies, create a dry, harsh surface more like sand than snow. Skis and sled runners, instead of melting the points of the crystals to make them slippery, merely roll the crystals over and over.

Snow changes continuously as it falls and after it has landed. Once on the ground, snowflakes trap tiny air pockets and form an excellent natural insulation. Temperatures on the surface can be more than 50 degrees colder than temperatures beneath seven inches of loose, fresh snow. As the snow settles it metamorphoses. "Old snow" is settled and dense, resulting from the altering of loose, pointed crystals into small round grains. Later it becomes firn, with spaces between the grains shrinking, resulting in compacting and hardening of snow. If metamorphosis continues long enough, firn can become glacial ice.

During any ordinary snowfall in New England or North Dakota or British Columbia or Siberia or Finland, about one million crystals of snow fall to cover each two-foot-square area with ten inches of snow. Snow covers approximately half the land on the earth's surface, at least for part of each year, as well as about 10 percent of the oceans. About 48 million square miles of the earth are covered year-round with snow or ice.

The greatest snowfall in a twenty-four-hour period recorded in North America occurred on April 14 and 15, 1921, when seventy-six inches of snow fell on Silver Lake, Colorado. More recently, on April 5 and 6, 1969, Bessans, France, was buried beneath sixty-eight inches of snow in nineteen hours. During a snowstorm from February 13 to 19, 1959, 189 inches fell on the Mount Shasta Ski Bowl in northern California. The snowiest place on record in North America is Rainier Paradise Ranger Station in Washington, where in the winter of 1971–72 a total of 1,122 inches of snow fell. The greatest depth of snow ever measured on the ground at one time in North America was 451 inches—over 37½ feet—at Tamarack, California, on March 11, 1911. Oddly enough, the interior of Antarctica receives very little snow. Most of the precipitation at that coldest spot on earth falls in the form of ice crystals, with an annual precipitation equal to less than two inches of water—only slightly more than falls on the Sahara Desert each year. The vast ice cap at the center of the continent grows, but only slowly, over millions of years.

"As pure as the snow," may not be the purest of metaphors. Snow, it seems, contains much more than just frozen moisture and air. In fact it contains enough nitrates, calcium, sulphate, and potassium picked up from dust and atmospheric gases to make it an important source of agricultural nutrients in many parts of the world. It also contains less savory traces of industrial pollution. When snow crystals form in air contaminated with sulphur dioxide the result is acid snow, which accumulates on the ground in the winter and releases highly acidic meltwater into rivers and lakes in the spring.

The best time to catch and observe snow crystals is when the temperature is moderately cold (about 25 degrees Fahrenheit is ideal), with no wind to throw the flakes against each other and break their points. The crystals fall individually, or more often, sticking together in loose clusters that fall apart into separate crystals when they land. Wear a dark jacket or carry a piece of dark-colored fabric stretched over cardboard, and after it has been acclimated to the outside temperature it will preserve even the finest, most delicate crystals until you have had time to examine them.

Most crystals are an eighth of an inch or less in diameter, much smaller than we are led to expect from the representation on Christmas cards. Mixed with those eighth-inchers are occasional midgets hardly larger than the dot on this letter i. Occasionally comes a behemoth, perfectly symmetrical and ornate as baroque jewelry, measuring as big around as a pencil eraser. Those giants spiral slowly downward, their flat surfaces horizontal, and are especially satisfying to catch on sleeve or tongue.

Nature's Tantrums: Blizzards

The storms of winter rarely begin with crash and clamor. They start quietly, growing by small degrees, engulfing us almost without notice. The first indication of a coming blizzard might be a slow drop in barometric pressure, accompanied by a thickening of cloud cover and a leaden appearance to the sky. Snow may or may not fall, but if it does the flakes are often small, hard ice pellets, rounded and packed by turbulence far up in the atmosphere, and will fall scooting before winds that do not seem —at first—strong enough to be alarming. The day before a winter storm might be unusually warm. Rain might fall. By evening a southerly breeze may have caused temperatures to rise high enough to give false hope of spring.

Then, during the night, the wind shifts to the north and the temperature drops, as much as 50 or 60 degrees. Snow that had been spitting erratically begins to fall more heavily, the flakes larger but still hard and rounded from battering one another as they descend. The wind gusts, picking up loose snow into sudden swirls, increasing gradually in intensity until there are no longer spaces between the gusts. Those of us who went to bed dreaming of birdsong and lilac blossoms wake to a world gone mad with whiteness and wind.

To qualify as a blizzard, meteorologists say a storm must contain the following characteristics: low or rapidly falling temperatures, winds of at least gale force (thirty-nine to forty-six miles per hour), and driving snow. Blizzards occur most frequently in the northern Great Plains, the prairie provinces of Canada, the upper Mississippi Valley, and the eastern Arctic.

Precipitation is not absolutely required for a blizzard. Loose, dry

snow can be picked up and blown along the ground in strong winds, creating all the conditions necessary for a blizzard. Relatively low winds create a moving layer of blown snow up to a few feet high, known as *low drift*. When stronger winds raise the layer of drifting snow above a human's head it is called *high drift*. One common result of all that blowing snow is what is known in casual usage as a *whiteout*. A more formal definition refers to a phenomenon of the polar ice caps, where reflections from the sky and snow create blinding glare that disorients a viewer. In such conditions depth perception becomes so uncertain that a mitten on the snow might look like a distant automobile.

For a blizzard to be worthy of the name, it must approach the intensity of winter storms on America's Great Plains. Consider this typical winter storm that struck South Dakota on January 22, 1952: In the middle and southern portions of eastern South Dakota, winds reached forty to fifty miles per hour, snow formed drifts ten to fifteen feet high, mail service was delayed for five days, eight human lives were lost, 2,500 cattle died, all primary roads were closed for three days, all side roads remained closed much longer, and it became necessary to airlift supplies to some regions. Temperatures on January 22 fell from a high of 21 degrees Fahrenheit to 0 degrees Fahrenheit by 7:00 a.m., to −5 degrees by 10:30 a.m., to −11 degrees by midnight; at its coldest, the windchill index reached approximately −75 degrees.

Not all serious blizzards are a Midwestern phenomenon. One of the worst in U.S. history came to be known as the Blizzard of '88 after it struck the eastern United States on Monday, March 12, 1888, and lasted through Wednesday, March 14, affecting one-quarter of the nation's population of the time and cutting off hundreds of cities from Maryland to Maine. The storm halted New York and over a dozen other major cities in their tracks, stopping virtually all transportation and communication. On that Monday, winds reached forty-eight miles per hour in New York City. Snowfall averaged forty to fifty inches over southern New England and southeastern New York State, with drifts thirty to forty feet high. Snowdrifts in Middletown, New York, covered three-story houses; townspeople tunneled through, shoring up the tunnels with timbers. At sea, where mariners called the storm the "Great White Hurricane," winds up to ninety miles per hour and waves up to three stories high were reported. At least 198 ships were lost—sunk or grounded—with the loss of about 100 seamen. On land, the death rate from freezing or storm-related accidents and illnesses approached 300, with 200 dead in New York City alone. For years people gathered on the anniversary of the blizzard to recall the storm. Others would never forget it, even if they wanted to: Dozens of babies born on or shortly after March 12 that year were named "Blizzard," "Storm," "Tempest," "Snowdrift," "Snowflake," and "Snowdrop."

Our word blizzard has controversial origins. After the Blizzard of '88 hammered the east coast of the United States then hopped the Atlantic and struck England, the London Times reported the word had its source in a common expression of the English Midlands, where "May I be blizzered" meant a person was "bowled over, or knocked off his feet"—in the manner of someone caught in a snowstorm. The New York Times responded indignantly that the word was of American origin: "The word is simply a bit of onomatopoeia. Like the hoof-beats in Virgil's poetry . . . the word is supposed to sound more or less like the thing it denotes."

When the storm struck Germany on March 19, 1888, it was referred to as the American Blizzard, and German newspapers reported the word originated from the German blitz. American author Gary Lockhart, in The Weather Companion, found the first printed mention of blizzard in the Northern Vindicator of Esterville, Iowa, in 1870: "Many of the early settlers in this area were from Germany, and when witnessing the severe winter storms, would use the German expression 'Der Sturm kommt blitzartig,' meaning 'the storm comes lightning-like.' The transition from blitzartig to blizzard was a natural language progression."

Among the notable blizzards in the United States was a 1717 storm on Cape Cod. In his diary, clergyman Cotton Mather described the storm as so severe that cottages were completely covered with snow, with "not the very tops of the Chimneys to be seen," and Indians 100 years old said "their Fathers never told

them of anything that equalled it." An unexpected storm in November 1846 trapped the seventy-nine men, women, and children
of the Donner party in a pass in the Sierra Nevadas; by the time
they were rescued in the spring, thirty-four had died. So many
blizzards swept across the Midwest during the "Massacre Winter"
of 1856–57, that the last of the prairie elk were wiped out in Iowa.
An 1867 storm in New York forced ferries to stop service across
the East River and was used as an argument for building the Brook-
lyn Bridge. In January 1888 a blizzard in Dakota Territory became
known as the "Schoolchildren's Storm," because it caused scores
of deaths while children were in school on a day that had begun
mild and sunny. In 1949 "The Blizzard of Forty-Nine" struck a
dozen western states and brought all ordinary routine to a stop;
states from Montana south to Arizona, and as far east as Kansas and
the Dakotas were paralyzed; snow covered the Mojave Desert,
stranded fifty trains between Omaha, Nebraska, and American
Falls, Idaho, and trapped 343 travelers in a tavern in Rockport,
Colorado, for three days.

Serious as those storms were, no place on earth experiences
blizzards that can compare to those of Antarctica. They usually
arise when winds from a low-pressure system combine with the
usual katabatic winds of the region. Wind speeds of ninety miles
per hour and gusts of 120 miles per hour are fairly common. Blizzards occur eight to ten times each year and can last for several
days. Perhaps the windiest place on earth—and consequently the
place with the most severe blizzards—is Cape Denison, Antarctica,
where an expedition led by the Australian explorer Douglas Mawson kept weather records during the winter of 1912. In March and
April the wind blew nearly continuously at sixty to eighty miles
per hour, with frequent gusts of 100 miles per hour, occasionally
reaching 200 miles per hour. In May, the average wind speed was
60.7 miles per hour, and for twenty-four hours on May 15 it averaged ninety miles per hour. For the entire year at Cape Denison
the average wind speed was nearly fifty miles per hour, about five
times the average in the United States or Europe. Even during lulls
in the wind, Mawson and his men were often caught in miniature

whirlwinds they called "whirlies," that were capable of picking up and hurling objects weighing hundreds of pounds. The entire winter amounted to one sustained blizzard.

In 1934 Richard Byrd, the famous navigator who was the first to fly over both poles, spent a winter alone in a hut in the Antarctic interior. He later wrote:

It is a queer experience to watch a blizzard rise. First there is the wind, rising out of nowhere. Then the Barrier [as the region where he stayed is known] unwrenches itself from quietude; and the surface, which just before had seemed as hard and polished as metal, begins to run like a making sea. Sometimes, if the wind strikes hard, the drift comes across the Barrier like a hurrying white cloud, tossed hundreds of feet in the air. Other times the growth is gradual. You become conscious of a general slithering movement on all sides. The air fills with tiny scraping and sliding and rustling sounds as the first loose crystals stir. In a little while they are moving as solidly as an incoming tide, which creams over the ankles, then surges to the waist, and finally is at the throat. I have walked in drift so thick as not to be able to see a foot ahead of me; yet, when I glanced up, I could see the stars shining through the thin layer just overhead. . . .

I didn't have any idea how really bad it was until I went aloft for an observation. As I pushed back the trapdoor, the drift met me like a moving wall. It was only a few steps from the ladder to the instrument shelter, but it seemed more like a mile. The air came at me in snowy rushes; I breasted it as I might a heavy surf. No night had ever seemed so dark. The beam from the flashlight was choked in its throat; I could not see my hand before my face. . . .

The noise was as if the entire physical world were tearing itself to pieces. I could scarcely heave the trapdoor open. The instant it came clear I was plunged into a blinding smother. I came out crawling, clinging to the handle of the door until I made sure of my bearings. Then I let the door fall shut, not wanting the tunnel filled with drift. To see was impossible. Millions of tiny pellets exploded in my eyes, stinging like BB shot. It was even hard to breathe, because snow instantly clogged the mouth and nostrils.

I made my way toward the anemometer pole on hands and knees, scared that I might be bowled off my feet if I stood erect; one false step and I should be lost forever.

To those of us who live in more temperate regions, secure in houses with dependable heating systems and cupboards well-stocked with food, blizzards are more cause for delight than alarm. Seen through the windows of a warm house, a snowstorm is a beautiful and awesome display. It awakens the child in us. We grow excited at the sight of the world transformed by snow and wind, at the sound of the wind howling outside, at the radio announcing the closing of schools and highways. A reprieve! The world calls time-out, and the responsibilities of adulthood are pushed aside for a day. It is the perfect excuse to remain in bed all morning, imitating the deep sleep of hibernation.

Raining Ice

ICE STORMS CAN DO more damage to wildlife, plants, and buildings than the swirling snow and high winds of blizzards. The conditions under which an ice storm can occur are strictly defined. Precipitation must begin as rain, and fall for a relatively short distance (the farther it falls the better the chance it will change to snow or hail). It must either be supercooled rain falling at less than freezing temperatures or a cold rain falling on a surface colder than the freezing point. Because the conditions necessary for an ice storm vary so little from the conditions of other winter storms, they are difficult to forecast. Generally, however, conditions are right when a warm front moves across a region of extreme cold and frost. Rain falling from the warmer air above will approach freezing as it nears the cold ground, then will freeze on contact, coating every earthbound surface with a layer of ice as much as three inches thick. The effect can be stunning and beautiful.

It can also be highly destructive. An ice storm that struck England in January 1940 was so severe that tele-graph poles snapped from the pressure of wires coated with up to 1,000 pounds of ice per span. Thousands of birds died of starvation and freezing when their feet froze to branches and wires. A severe ice storm, triggered by a massive polar front sweeping down the Mississippi Valley from Canada, struck the central United States in January 1951. From January 28 to February 1, freezing rain totaling more than five inches of precipitation fell on parts of Tennessee, Kentucky, and West Virginia, blanketing everything with a dense coating of ice, breaking down trees, power lines, and telephone lines, collaps-ing roofs, and stranding thousands of people without power, transportation, or communication for more than a week.

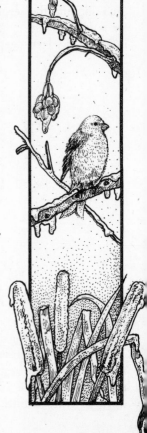

In North America, ice storms are at least an occasional hazard everywhere except the Yukon, Northwest Territories, and the high arctic regions of Canada, where it is usually too cold for freezing rain; southern California, Nevada, Utah, Arizona, New Mexico, where it is generally too dry; and Florida, where it is generally too warm.

Damage is usually limited to trees, especially large, old trees with prominent Y-forks, and there is little that can be done to protect them except to wire weak forks together. Many botanists say don't bother: Freezing rain is nature's pruning saw, clearing old branches to make way for new growth.

TOUGH BIRDS

Endure northern winters long enough and you are sure to become fascinated with migration. My neighbors who spend winters in Florida or Arizona, and who are known by the natives as "snowbirds," think of themselves as yet another winged species escaping the inhospitable north. In the fall, those of us who stay put—ornithologists would label us residents as opposed to migrators—are glad we are not leaving. By midwinter, we become prone to a certain wistfulness. Why have we and the chickadees decided to stay? If migration works to satisfaction for some birds—and it does for most—why not all?

Such questions are not easily answered. Many species resist easy classification as resident or migratory. Some birds—blue jays, for instance—migrate one year and not the next, or migrate such short distances you wonder why they bother. Others are irruptive, migrating only during those years when food is scarce. Still others, like dark-eyed juncos in North America and chaffinches in northern Europe, are partial migrators, with often the dominant individuals remaining in the north while smaller individuals migrate south where there is less competition for food.

Availability of food is the primary criterion determining where birds spend the winter. Flycatchers, swallows, nightjars, and other species that depend almost entirely on airborne insects for their diet are forced to migrate south when autumn weather begins to kill flying bugs. Wading birds and waterfowl are imperiled by frozen water. Raptors dependent on warmth-induced air currents to keep them aloft while hunting move south as cold deadens the air. Some years the northern United States are inundated by large numbers of snowy owls, which will be frequently seen perched on fence posts and telephone poles. According to folk knowledge, sightings of snowy owls indicated a particularly hard winter to

301

come, since the birds must have been driven south of their customary habitat in Canada by extremely cold weather. In truth, however, they periodically expand their range to the south in search of food. Their primary item of diet is the Arctic lemming, which has a fluctuating population cycle. When there are not enough lemmings to support the snowy owls, they move south in search of other prey.

Those birds that remain in snow-covered country in winter—and only twenty to thirty species ordinarily do so in North America—must be able to live off what food they can find. Fall berries and seeds sustain sparrows, finches, cardinals, and grosbeaks. Scavengers such as gulls, pigeons, crows, and ravens rely on everything from unharvested field crops to roadkill to garbage. About half the overwintering birds, including nuthatches, chickadees, titmice, kinglets, brown creepers, and woodpeckers, are equipped to extract hibernating insects and their larvae and eggs from nooks and burrows in trees; many will supplement that diet with seeds, nuts, and berries. A few species make efforts to store food caches. A blue jay is a diligent collector of seeds and nuts during autumn, hiding them in crevices in trees and under litter on the ground, but jays usually lose what they stockpile, or are overcome with absent-mindedness and eat an entire day's hoarding in a few minutes. The gray jay, also known in various parts of the northern states and Canada as the whisky-jack, Canada jay, or, aptly, camp robber—for

its willingness to steal food right out of your Coleman cooler—is more skilled, using salivary excretions to form sticky lumps of food called boluses, which it stores on twigs and conifer needles. Nuthatches and titmice stockpile seeds in tree bark, then often camouflage their caches by covering them with bits of lichen and bark.

Besides finding enough food, birds rely on a number of tricks to beat the challenge of winter's cold weather. One way they do that is to increase their body fat. Nonmigratory dark-eyed juncos in Michigan were found to weigh 14 percent more in winter than the same species in Alabama, yet even that much extra fat is only enough reserve to last sixteen to twenty-four hours—not much more than one cold night. Those birds that can store and maintain the most fat have the best chance of surviving.

Body size is another important factor in winter survival. Large birds, because they have a greater ratio of mass to surface area, are able to conserve energy longer and more efficiently. Natural selection seems to favor the concept, known as Bergmann's Rule, leading to significant size differences within species widely divided by geography. The hairy woodpecker of Canada, for instance, is considerably larger than the same species in Costa Rica, and the emperor penguins of Antarctica dwarf their cousins in the Galapagos and Australia.

At night most birds seek only the minimal shelter afforded by dense trees and shrubs, but a few species, especially pygmy nuthatches, wrens, and brown creepers, increase their chances of living through cold nights by huddling together in protected roosts. The ornithological journal *Condor* in 1957 reported that 100 nuthatches were discovered roosting so closely together in a pine tree cavity that some had suffocated. A more recent study of kinglets found that two birds huddling cut heat loss by one-quarter, while three saved one-third.

Ruffed grouse, willow ptarmigan, willow tits, Siberian tits, and common redpolls conserve body heat by burrowing into loose snow to take advantage of its natural insulation. In a sheltered roost beneath eight or ten inches of snow, where the temperature remains as much as 40 degrees warmer than outside air, energy loss can be reduced by nearly half.

Black-capped chickadees are successful overwintering birds, a remarkable fact considering their tiny size. Such a small bird, with so little body mass, cools off much faster than larger birds and must work harder to keep fueled and warm. Yet chickadees go through the day in such obvious good spirits it is difficult not to be cheered by them. They are able somehow to look both amused and curious as they cock their heads to examine a bit of tree bark or watch the odd activities of a human shoveling a walk near the bird feeder. As they go about the business of finding food they seem absolutely indifferent to the potentially lethal wind.

They survive that wind by employing a number of tactics to conserve heat. Chickadees and other small passerines, equipped with the highest metabolic rate of any vertebrates and having less body mass in relation to surface area than larger birds, must take every possible precaution to maintain their 104-degree Fahrenheit body temperature. Their first line of defense is to reduce the rate of heat loss. Like most birds, they employ a tactic known as pilo-erection, fluffing their feathers to trap air and enhance insulation value. They also constrict surface blood vessels to reduce the circulation of blood in extremities like feet and legs, and take shelter whenever possible. On the coldest days they hide in dense vegetation or in the hollows formed in snow-covered shrubs. At night, they are able to descend into a state of mild hypothermia, or torpor, lowering their body temperatures gradually by as much as 20 degrees Fahrenheit, reducing heat flow to the outside air and saving about 20 percent of the energy they would otherwise burn.

While such methods of heat conservation are critical to winter survival, they would probably be futile without a way to produce and regulate additional heat. Most birds, chickadees included, solve that problem by shivering. When not generating heat through the muscle activity of flight, small birds (and perhaps larger species as well) are thought to shiver almost constantly during cold weather. The muscle contractions of shivering increases the rate of metabolism as much as five-fold, making it possible to generate high levels of heat for long periods of time.

Shivering is something we overwintering species understand.

Late on a cold January afternoon I stand in my yard watching blue jays plunder the bird feeder. Studies suggest that the feeders I keep valiantly stocked all winter are not terribly important to the survival of wintering birds, but I like to think I'm doing my small part to help. The jays, searching impatiently for the fattest sunflower seeds, scatter lesser seeds onto the snow below. A half-dozen goldfinches perch in the branches of a forsythia nearby, waiting for the jays to leave. The finches are shivering. The jays are shivering. I'm shivering. For the moment we're all getting by just fine.

SELECTED BIBLIOGRAPHY

Astronomy and Cosmology

Asimov, Isaac. 1985. *Asimov's Guide to Halley's Comet.* New York: Dell Publishing Co.

Brown, Peter Lancaster. 1990. *Star and Planet Spotting: A Field Guide to the Night Sky.* New York: Sterling Publishing Co.

Cavendish, Richard, ed.-in-chief. 1985. *Man, Myth and Magic: The Illustrated Encyclopedia of Mythology, Religion and the Unknown.* New York: Marshall Cavendish.

Corliss, William R. 1979. *Mysterious Universe: A Handbook of Astronomical Anomalies.* Glen Arm, Maryland: The Sourcebook Project.

Dodd, Robert T. 1986. *Thunderstones and Shooting Stars: The Meaning of Meteorites.* Cambridge: Harvard University Press.

Durham, Frank, and Robert D. Purrington. 1983. *Frame of the Universe: A History of Physical Cosmology.* New York: Columbia University Press.

Etter, Roberta, and Stuart Schneider. 1985. *Halley's Comet: Memories of 1910.* New York: Abbeville Press.

Flaste, Richard, Holcomb Noble, Walter Sullivan, and John Noble Wilford. 1985. *The New York Times Guide to the Return of Halley's Comet.* New York: Times Books.

Harrington, Philip S. 1990. *Touring the Universe through Binoculars.* New York: John Wiley & Sons.

Kals, W. S. 1990. *Stars and Planets: The Sierra Club Guide to Sky Watching and Direction Finding.* San Francisco: Sierra Club Books.

Krupp, E. C. 1991. *Beyond the Blue Horizon: Myths and Legends of the Sun, Moon, Stars, and Planets.* New York: HarperCollins.

Leiber, Arnold L., M.D. 1988. *The Lunar Effect: Biological Tides and Human Emotions.* Garden City, New York: Anchor Press/Doubleday.

Levy, David H. 1991. *The Sky: A User's Guide.* New York: Cambridge University Press.

Menzel, Donald H., and Jay M. Pasachoff. 1983. *A Field Guide to the Stars and Planets.* Boston: Peterson Field Guide Series, Houghton Mifflin.

Moore, Patrick. 1968. *Suns, Myths and Men.* New York: W.W. Norton and Company.

Price, Fred W. 1988. *The Moon Observer's Handbook.* New York: Cambridge University Press.

Ronan, Colin A. 1985. *The Skywatcher's Handbook.* New York: Crown Publishers.

Sagan, Carl. 1980. *Cosmos.* New York: Random House.

Schaaf, Fred. 1988. *The Starry Room: Naked Eye Astronomy in the Intimate Universe.* New York: John Wiley & Sons.

Trefil, James S. 1985. *Space, Time, Infinity: The Smithsonian Views the Universe.* New York: Pantheon.

Wentzel, Donat G. 1989. *The Restless Sun.* Washington, D.C.: Smithsonian Institution Press.

Wilson, Colin. 1980. *Starseekers.* New York: Doubleday and Company.

Yeomans, Donald K. 1991. *Comets: A Chronological History of Observation, Science, Myth, and Folklore.* New York: John Wiley & Sons.

Meteorology

Allen, Oliver E. 1983. *Atmosphere.* Alexandria, Virginia: Time-Life Books.

Anderson, Bette Roda. 1975. *Weather in the West.* Palo Alto, California: American West Publishing Co.

Aristotle. *Meteorologica.* Harvard University Press edition, Cambridge, 1962.

Blanchard, Duncan C. 1967. *From Raindrops to Volcanoes.* Garden City, N.Y.: Doubleday and Company.

Burroughs, William James. 1991. *Watching the World's Weather.* New York: Cambridge University Press.

Cable, Mary. 1988. *The Blizzard of '88.* New York: Atheneum.

Calder, Nigel. 1975. *The Weather Machine.* New York: Viking.

Cotton, William R. 1990. *Storms.* Fort Collins, Colorado: Geophysical Science Series Volume 1, Aster Press.

de la Rue, E. Aubert. 1955. *Man and the Winds.* New York: Hutchinson and Co.

Erickson, Jon. 1988. *Violent Storms.* Blue Ridge Summit, PA: TAB Books.

Frazier, Kendrick. 1979. *The Violent Face of Nature: Severe Phenomena and Natural Disasters.* New York: William Morrow and Company.

Freier, George D. 1989. *Weather Proverbs.* Tucson, Arizona: Fisher Books.

Hughes, Patrick. 1976. *American Weather Stories.* Washington, D.C.: U.S. Department of Commerce.

Kimble, George H. T. 1955. *Our American Weather*. New York: McGraw-Hill.

Kirk, Ruth. 1977. *Snow*. New York: William Morrow.

Kotsch, William J. 1983. *Weather for the Mariner*. Annapolis, Maryland: Naval Institute Press.

Lee, Albert. 1976. *Weather Wisdom: Facts and Folklore of Weather Forecasting*. Chicago: Congdon and Weed.

Lockhart, Gary. 1988. *The Weather Companion: An Album of Meteorological History, Science, Legend, and Folklore*. New York: John Wiley & Sons.

Meinel, Aden, and Marjorie Meinel. 1983. *Sunsets, Twilights, and Evening Skies*. New York: Cambridge University Press.

Neider, Charles, ed. 1972. *Antarctica*. New York: Random House.

Parfit, Michael. 1985. *South Light, A Journey to the Last Continent*. New York: Macmillan.

Schaaf, Fred. 1983. *Wonders of the Sky: Observing Rainbows, Comets, Eclipses, the Stars, and Other Phenomena*. New York: Dover Publications.

Schaefer, Vincent J., and John A. Day. 1981. *A Field Guide to the Atmosphere*. Boston: Houghton Mifflin Company.

Schonland, Sir Basil. 1964. *The Flight of Thunderbolts*. London: Oxford University Press.

Scorer, Richard, and Arjen Verkaik. 1989. *Spacious Skies*. London: David and Charles.

Watson, Lyall. 1984. *Heaven's Breath: A Natural History of the Wind*. New York: William Morrow and Company.

Young, Louise B. 1977. *Earth's Aura*. New York: Knopf.

Nature, General

Ackerman, Diane. 1990. *A Natural History of the Senses*. New York: Random House.

Asimov, Isaac. 1987. *Beginning: The Story of Origins—of Mankind, Life, the Earth, the Universe*. New York: Walker and Company.

Attenborough, David. 1979. *Life on Earth*. Boston: Little, Brown and Company.

Ayensu, Edward S., and Philip Whitfield, consultant editors. 1982. *The Rhythms of Life*. New York: Crown Publishers.

Borland, Hal. 1973. *Seasons*. New York: Lippincott.

Ehrlich, Paul R. 1986. *The Machinery of Nature*. New York: Simon and Schuster.

Fuchs, Sir Vivian, ed. 1977. *Forces of Nature*. New York: Holt, Rinehart and Winston.

Gray, William, ed. 1978. *Powers of Nature*. Washington, D.C.: National Geographic Society.

Halfpenny, James C., and Roy Douglas Ozanne. 1989. *Winter: An Ecological Handbook*. Boulder, Colorado: Johnson Publishing Company.

Krutch, Joseph Wood. 1972. *The Voice of the Desert*. New York: William Sloan Associates.

Leopold, Aldo. 1949. *A Sand County Almanac*. New York: Oxford University Press.

Lucretius. *The Nature of the Universe*. Baltimore: Penguin Books edition, 1965.

Platt, Rutherford. 1971. *Water: The Wonder of Life*. Englewood Cliffs, N.J.: Prentice-Hall, Inc.

Pliny. *Natural History*. Cambridge: Harvard University Press edition, 1958.

Teale, Edwin Way. 1965. *Wandering Through Winter*. New York: Dodd, Mead & Company.

———. 1960. *Journey Into Summer*. New York: Dodd, Mead & Company.

———. 1956. *Autumn Across America*. New York: Dodd, Mead & Company.

———. 1951. *North With the Spring*. New York: Dodd, Mead & Company.

Nature, Wildlife

Bent, Arthur Cleveland. 1961. *Life Histories of North American Birds of Prey, Part One*. New York: Dover Publications.

Berenbaum, May R. 1989. *Ninety-nine Gnats, Nits, and Nibblers*. Chicago: University of Illinois Press.

Broley, Myrtle Jeanne. 1952. *Eagle Man: Charles L. Broley's Field Adventures with American Eagles*. New York: Pellegrini and Cudahy Publishers.

Brooks, Bruce. 1989. *On the Wing*. New York: Charles Scribner's Sons.

Burton, Robert. 1990. *Bird Flight*. New York: Facts on File.

Cloudsley-Thompson, J. L. 1965. *Desert Life*. London: Pergamon Press.

Ehrlich, Paul R., David S. Dobkin, and Darryl Wheye. 1988. *The Birder's Handbook: A Field Guide to the Natural History of North American Birds*. New York: Simon and Schuster.

d'Entreves, P. Passarin, and M. Zunino. 1976. *The Secret Life of Insects*. London: Orbis Publishing Limited.

Evans, Howard Ensign. *Life on a Little-known Planet.* 1968. New York: E.P. Dutton and Co., Inc.

Fabre, J. Henri. *The Insect World of J. Henri Fabre.* Boston: Beacon Press edition, 1991.

George, Jean Craighead. 1970. *Beastly Inventions.* New York: David McKay Co.

Gerrard, Jon M., and Gary R. Bortolotti. 1988. *The Bald Eagle: Haunts and Habits of a Wilderness Monarch.* Washington, D.C.: Smithsonian Institution Press.

Gill, Frank B. 1990. *Ornithology.* New York: W. H. Freeman and Co.

Hanson, Jeanne K., and Deane Morrison. 1991. *Of Kinkajous, Capybaras, Horned Beetles, Seladangs, and the Oddest and Most Wonderful Mammals, Insects, Birds, and Plants of Our World.* New York: HarperCollins.

Heinrich, Bernd. 1984. *In a Patch of Fireweed.* Cambridge, Mass.: Harvard University Press.

Irving, L. 1972. *Arctic Life of Birds and Mammals.* New York: Springer-Verlag.

Klausnitzer, Bernhard. 1987. *Insects: Their Biology and Cultural History.* New York: Universe Books.

Linsenmaier, Walter. 1972. *Insects of the World.* New York: McGraw-Hill.

Marchand, Peter J. 1987. *Life in the Cold: An Introduction to Winter Ecology.* Hanover, New Hampshire: University Press of New England.

Milne, Lorus, and Margery Milne. 1962. *The Senses of Animals and Men.* New York: Atheneum.

Perrins, Christopher M., consultant-in-chief. 1990. *The Illustrated Encyclopedia of Birds.* New York: Prentice-Hall.

Root, Terry. 1988. *Atlas of Wintering North American Birds, An Analysis of Christmas Bird Count Data.* Chicago: University of Chicago Press.

Schmidt-Nielsen, Knut. 1964. *Desert Animals: Physiological Problems of Heat and Water.* London: Oxford University Press.

Stokes, Donald W. 1983. *A Guide to Observing Insect Lives.* Boston: Little, Brown and Company.

Science, General

Asimov, Isaac. 1989. *Asimov's Chronology of Science and Discovery.* New York: Harper and Row.

Burke, James. 1985. *The Day the Universe Changed.* Boston: Little, Brown and Company.

Flatow, Ira. 1988. *Rainbows, Curve Balls, and Other Wonders of the Natural World Explained*. New York: William Morrow and Company.

Hazen, Robert M., and James Trefil. 1991. *Science Matters: Achieving Scientific Literacy*. New York: Doubleday.

Trefil, James. 1989. *Reading the Mind of God*. New York: Charles Scribner's Sons.

———. 1987. *Meditations at Sunset*. New York: Charles Scribner's Sons.

von Baeyer, Hans C. 1984. *Rainbows, Snowflakes, and Quarks*. New York: McGraw-Hill.

INDEX

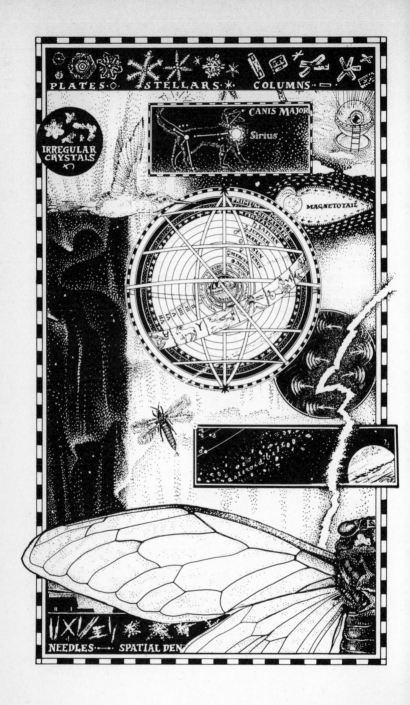

PLATES · STELLARS · COLUMNS

CANIS MAJOR
Sirius

IRREGULAR CRYSTALS

MAGNETOTAIL

PRIMUM MOBILE

asteroid belt

NEEDLES ⟶ SPATIAL DEN

ABOUT THE AUTHOR
AND ILLUSTRATOR

Jerry Dennis has traveled extensively pursuing his interest in outdoor subjects. He has contributed articles to the *New York Times*, *Audubon*, *Country Journal*, *Sports Afield*, *Outdoor Life*, *Canoe*, and *The Utne Reader*, among other publications. His column on unusual wildlife behavior, "The Natural Enquirer," appears in each issue of *Wildlife Conservation*. He is the coauthor of *The Best Bicycle Tours of Eastern Canada*. He lives in northern Michigan with his wife and two sons.

Glenn Wolff's illustrations have appeared in the *New York Times*, *Sports Afield*, *Sports Illustrated*, *Audubon*, the Bulletin of the Atomic Scientists, the *Village Voice*, *The Progressive*, and in "The Natural Enquirer" column in *Wildlife Conservation*. He has also illustrated more than a dozen books. He lives with his wife and two daughters in northern Michigan.